John Glad

Future Human Evolution

Eugenics in the Twenty-First Century

Preface by Seymour W. Itzkoff

Hermitage Publishers
2006

John Glad
FUTURE HUMAN EVOLUTION
Eugenics in the Twenty-First Century

Photography by Richard Robin

Excerpts from this book have appeared in *Mankind Quarterly* and *Jewish Press.*

Library of Congress Cataloging-in-Publication Data:

Glad, John.
Future human evolution: eugenics in the twenty-first century / John Glad.
p. cm.
Includes bibliographical references and index.
ISBN 1-55779-154-6
1. Eugenics. I. Title.
HQ751.G52 2005
363.9'2—dc22
2005052536

Published by Hermitage Publishers
P.O. Box 578
Schuylkill Haven, PA 17972-0578
E-mail: yefimovim@aol.com
The entire Hermitage Publishers catalog is available on the Internet:
www.Hermitagepublishers.com

This book may be downloaded free of charge at www.whatwemaybe.org.

Acknowlegements

I wish to express my gratitude to all those who gave so generously of their time in preparing the various drafts of this book: Carl Bajema, Norman DiGiovanni, Sarah Forman, Larisa Glad, Oleg Panczenko, Richard Robin, Alex Van Oss, James Woodbury, and Ilya Zakharov.

Table of Contents

Preface

John Glad is a brave scholar. He here ventures onto the high seas of contemporary intellectual *interdict*. The term *eugenics* has been on an ideological hit list both by the irrational left as well as by an intimidated public. However, as Dr. Glad points out, clearly and authoritatively, there is virtually no factual basis for what can only be seen as a totemic reaction. The mere mention of *eugenics* elicits knee-jerk reaction—"Nazi genocide, forced sterilization." Yet by any standard of rational analysis, this vision of improvement for the human species has a strong humanistic tradition to support its further application.

The real history of *eugenics,* as Dr. Glad points, out is rich in a truly liberal vision for the improvement in the state of all of humankind. And modern research in the biological nature of human function is opening up opportunities for the enhancement of both the physical as well as the mental condition of the human species. This, at a blazing speed of discovery. Thus, we need thinkers such as John Glad who will step up to challenge blind prejudice with fact and possibility.

The world is in a descending spiral today, with 6.5 billion people, going on 9-10 billion humans by mid-century, the vast majority living under historically and civilizationally subhuman conditions.

The same powers-that-be, those that blind the educated with a fear of the term *eugenics,* represent the self-same leadership classes that benefit from the present futile redistributionist social policies that feed into the demographic explosion of the destitute and the vulnerable. What is occurring, and against which Dr. Glad is expostulating, is a shakedown and intimidation of the productive middle classes in order to feed the pathology of poverty, disease, and social disintegration to which we are exposed in the media, each day.

These ideological leadership cadres that stand in the way of the dissemination of the truth concerning the ideals of the old and new eugenics movement indulge themselves luxuriously in the watering places of the "philanthropists," in Paris,

Geneva, New York, Brussels. These international organizations—we know them well—fritter away billions of dollars for their own partying (they call them conferences), the remnant dollars dribbling supposedly into the lands of the needy, but really sucked up by the gangsters who run the tragic show of the Third World. The poor get poorer, their conditions of life increasingly pathological, unprecedented in scope at any time in history.

Eugenics, a vision of human betterment, with real scientific and then social-policy potential for enhancing the evolutionary future of our species, is buried within a demonization of language and misunderstanding. Critical to the linguistic and semantic morass that surrounds this paralysis of understanding is the spectral memories of the German and European perpetration of the *Holocaust*.

I would like to add a comment to Dr. Glad's clear and decisive puncturing of the balloon of myth that argues that the Nazis claimed to have actually engaged in a program of eugenics. The Nazis also claimed to be a party of *socialism*! If we define eugenics as encompassing programs of human betterment, physical as well as mental, practices that benefit community in the local sense as well as the species in general, we can say that the *Holocaust* was the antithesis of eugenic practice. Not only did the Nazis not argue for their participation in the eugenics movement, but they knew that they were practicing *dysgenics.*

They hid their practices, as do all totalitarian regimes, within a babble of propaganda that presumably validated to the naïve, this mirage of self-justification. A careful reading of their mission statements, and, of course, their unspeakable practices, clearly reveals that that they recognized that they were eliminating a people who they knew to be superior to themselves, a millennial threat to German dominance. They covered these actions by heaping slime on the Jewish people, their racial heritage, their ghetto and post-ghetto cultural behavior, their arrogance and purported economic conspiracies, above all their dominance in all walks of life, quickly attained only a brief moment beyond the ghetto. To the Nazis, this became a universal challenge to German pretensions to leadership. And this from a people that in Germany was a scant one percent of the population, in the entire Austro-Hungarian Empire, about four percent.

One has only to read the literature of polemics arising from the German/Austrian political/cultural scene, from the mid-nineteenth century on, to realize that the hatred of the Jews was not a hatred of religion, but rather of race. The solution, clearly and early bandied about by a wide variety of European hate groups, was one of potential cleansing of the Jews from Europe, if not the world. Simply, the polemics of hate was engendered to facilitate the elimination of a dangerous contender for dominance in this self-same continental environment.

Thus the genocide of the Jews, in which all of Europe became eager participants, was not an example of eugenics gone astray, as Dr. Glad suggests. I here, gently demur. Rather, the *Holocaust* was a vast dysgenic program to rid Europe of superior intelligent challengers to the existing Christian domination by a numerically and politically minuscule minority.

The issue of gypsy genocide has been continuously presented to throw dust in the air, to obfuscate the real significance of the fate of the Jews in Europe between 1933 and 1945. True, the gypsies were persecuted and Hitler disdained them. Yet the ethnic gypsies, as distinct from West European converts, represented, to the perverse irrationality of the Nazis, an ancient Aryan race. Thus, as Aryans, the gypsies were not subjected to premeditated total genocide

The genocide began with the Nazi accession to power in Germany, 1933; in Austria, 1938. It was both chaotic and bestial, but many German and Austrian Jews made good their escapes. There was truly hatred, a chaos of despicable cruelty in Germany, Austria, and the occupied lands up to January 1942, when the Nazis realized that Britain and the Soviet Union still stood strong against their aggression, while the United States, bruised after Pearl Harbor, rearmed in fury. At Wannsee, north of Berlin, the final solution was conjured up, the industrial annihilation of the remaining Jews of Europe. If Germany would not prevail, no Jews would be left to gloat vindictively of their own victory.

Another sad mental block over the real meaning of the *Holocaust*, and here within the Jewish community itself, is the Jews' refusal to accept this event as an exemplar of *dysgenics*. To do so, many fear, would only reify the view that the Jewish people still considered themselves among the *elect*, the *chosen*, as the *Torah* implies. To admit this would presumably again bring down a vale of tears upon them.

The events in Europe during these decades was thus not an exemplification of the theory of eugenics, a supposed liberal and humanitarian vision turned to dross. Rather it was, as noted above, a premeditated program of dysgenics, an *aristocide*, as with too many other genocides of the twentieth century. How else can we understand the ideology of hate during this century that brought about the destruction of so many talented human beings, members of civilizationally achieving ethnic and social class groups? Thus we have here witnessed, from Armenia to Biafra and Cambodia, the dysgenic destruction of tens of millions of the most intelligent, productive humans on our planet.

By not recognizing the twentieth century's true "achievement," we have thus given in to the defamation of the ideals of the eugenics movement. We have made far more difficult the wider clarification of the true implications of eugenics.

It is doubly important to emphasize the visionary qualities of Dr. Glad's book. Because, even after throwing over this contemptible myth of "Nazi eugenics," a twenty-first century campaign for the eugenic ideal must impress upon educated and uneducated alike that the problems that we face require a healthy humanity living in tune with nature. It requires a revolutionary turnabout from present dogmatic international thinking. Instead of dissipating our wealth to remediate what cannot be remediated we need to envision clearly what measures humanity needs to take to create a future of hope. Dr. Glad makes this clear: universal high intelligence, altruism, a pragmatic analysis of the facts of our current situation. Our world simply is running aground in majoritarian incapacity and with this impotence, potential medical and ecological disaster. What a program of eugenics offers potentially goes far beyond even the ongoing strong eugenic decisions made by millions of families with regard to procreation and the raising of healthy youngsters. Here, individuals, if not the power brokers, are obeying the laws of science and thereby acting to prevent more misery and suffering. What a programmatic campaign for eugenics on a worldwide basis could do over the decades if not centuries is to lift a curtain of hope, to be substituted for the cloud of concern that the middle classes have pessimistically internalized over the last decades.

We are on the cusp of a scientific reality, the uncovering of a human biological nature as never dreamed possible before. Not merely the identification of potential disabilities in unborn children, the solving of the sadness of infertility, even to the extent of cloning a desired child when no other pathway of biological reproduction is possible. Scientists today are, in addition, and all over the world, searching for enzymatic indicators during the earliest stages of gestation, for the genes of high and low intelligence. When these markers are discovered, given the acknowledged random nature of intelligence variability even within families, it will allow mothers and fathers to choose the potential intelligence of their child-to-be. The masses will here no doubt once more vote with their test tubes for a eugenic solution.

It may have been biologist Bentley Glass who once commented, eventually sexual relations would be freed from their reproductive role. Eugenics?

The rub is that we now have to teach the elites that biologically determinant decisions guided by scientific knowledge and careful judicial and moral monitoring can give us the world for which we yearn. Here is real, empirical, scientifically-supported evidence for humanity's hope, not the tragic morass of pathologies that the so-called egalitarians are pulling down over the heads of our grandchildren.

John Glad's *Future Human Evolution* is an important book. It needs many readers. I am sure it will achieve this goal.

Seymour W. Itzkoff

Introduction

I am with you, you men and women of a generation,
or ever so many generations hence.
Walt Whitman, "Crossing Brooklyn Ferry"

The Great War and subsequent Depression undermined the mentality of Empire and class privilege, leaving a vacuum which was filled by an intellectual climate of extreme egalitarianism. Western society of the twentieth century came to be dominated by a new, unified ideology. Freudianism, Marxism, B. F. Skinner's Behaviorism, Franz Boaz's cultural history, and Margaret Mead's anthropology all stressed the marvelous "plasticity" and even "programmability" of *Homo sapiens.* It was explained over and over that human minds differ little in their innate qualities, and that it is upbringing and education which explain the differences among us. Software is everything; hardware is identical and thus meaningless. The road to utopia lies through improved nurture alone.

During the last third of the twentieth century, even while scientists were generally allowed to teach the theory of evolution, that freedom did not extend to raising the topic of humanity's future evolution. It is remarkable that this suppression coincided with a revolution in our understanding of genetics. The censorship has now been lifted, and there is agreement even among the most implacable foes of the eugenics movement that the taboo on eugenics can no longer stand.

The issues involved are so fraught with consequence at all levels that, tiny as the group of individuals concerned over the future genetic composition of humankind is, a single ideological spark in this area has the potential to set off an all-consuming conflagration, so that hostility all too often squeezes out rational discussion. But no matter how desperately society attempts to avoid these issues, they already stand before us, demanding at least recognition, if not resolution. In this book I attempt to present the heretofore largely suppressed arguments surrounding the current renaissance of the eugenics movement.

*

Much as we humans might pride ourselves on our achievements, we are really little closer to resolving the great questions of being than when we still dwelled in caves. Time extending endlessly backward or forward is as unimaginable as is time having a beginning or an end. Psychologically, however, we need a map – a concept of being and of our place in the universe – and thus we engage in elaborate myth-making to fill the vacuum that we find so intolerable. To be durable, a worldview must first explain the universe to us, and then assuage our fears and satisfy our longings. Logic is not a prerequisite. Myth can even contradict itself – not to mention be at variance with the real world.

Regardless of when or where we live, we inevitably perceive ourselves as the Middle Kingdom, and either we smile condescendingly at the mythmaking of other cultures or we go to war with them to force upon them our (uniquely correct) worldview. And if we are better at crafting weapons, we are generally able to persuade those we have physically conquered of the superiority of our myths over theirs.

Until the mid-nineteenth century, the Western world accepted a literal interpretation of the Book of Genesis, but then the theory of evolution presented a radically different explanation of man's origins. Today, attempting to reconcile religion with science, we have created a new mythology which, not surprisingly, is ripe with contradictions. Here are some of them:

a) While other species of animal and plant can undergo significant change over a few generations, we maintain that thousands of generations of the most radically varying conditions of selection and selective mating have left only the most superficial genetic variance within our species.

b) Intellectuals (albeit not the man in the street) are firmly convinced that we are the product of evolution, but they are equally entrenched in the odd assumption that human beings are the one species no longer affected by that process.

c) Even as society pays a premium for ability and gumption in virtually any form of activity, it has become fashionable to claim that such factors play no role in the formation of social classes, which are held to be entirely a function of chance and privilege. Indeed, the scholars who dominate the publishing marketplace and academia deny the very existence of innate IQ variance in human populations.
d) We have developed a huge academic testing industry, but its findings are widely declared to be not merely approximate but lacking in any validity whatever.
e) With the transition to smaller families, we have observed that generation after generation of the intellectually endowed are failing to replace themselves– exactly as was feared by earlier eugenicists – but we accept the phenomenon as natural.
f) We are more and more successfully implementing a process called "medicine" for the elimination of natural selection, and are firmly convinced that future generations will remain unaffected by our reluctance to implement a substitute for natural selection.
g) Hard at work deciphering the map of the human genome, we continue to apply moral criteria to behavior which we will soon be able to explain scientifically.
h) While our social conduct, like that of all other animal species, is necessarily centered around the mating ritual, our perception of this process is governed by a myriad of camouflaging taboos and fetishes. The gap between reality and fantasy could not be more crass.
i) We have created a genetic caste society that co-opts talent born into the less privileged castes, efficiently exploiting and manipulating these castes, while at the same time proclaiming equality of opportunity as our slogan.

j) We refuse to recognize that we are a species that perfectly fits the definition of a disease, freeing itself (very temporarily) from the constraints of natural selection and the limitations of natural resources only to wreak havoc on ourselves and our fellow species in a massive assault on the host that we parasitize – the planet.
k) We have created an unsustainable economy dependent on resource exhaustion. At the same time, we proclaim still greater levels of consumption as the goal of society.
l) We proclaim freedom of speech, all the while ruthlessly excoriating any opinion in the area of human genetics which is found offensive by any significant segment of society.

Thus, the revolution in technology has been accompanied, not by the elimination of myth, but by its modification into a denial of biology. The give and take of any political processes is necessarily determined by the relative power of the participants, so that future generations are not taken into consideration during decision-making.

Despite popular opinion and prejudice, the facts of science are inescapable. In the time you take to read this sentence, humankind will have evolved genetically. There are species such as the coelacanth fish, which – incredibly – has survived more than 400 million years, but they are the rare exception. *Homo sapiens* is a recent link in the evolutionary chain, and over the past century the conditions governing selection in that population have undergone revolutionary changes.

Ultimately, we have to decide how pleased we are with ourselves as a species. This is the great watershed dividing those who favor genetic intervention and those who oppose it. Regardless of our personal attitudes, however, there is no denying the fact that while the genetic lottery has indeed produced many winners, there are many others who have been less fortunate.

The eugenics movement, which can be understood as human ecology, has long considered itself a lobby for future generations, arguing that while it is true that we should not be presumptuous in our ability to predict the future, we can define what we want – healthy, intelligent babies who will grow up to be emotionally balanced, broadly altruistic adults.

Now, when the majority of people live far beyond their child-bearing years, it is not those who have survived a horrendous process of natural selection who will populate the planet in the future, but those who have the most offspring. We now have selection by fertility rather than by mortality – a revolutionary change.

On a theoretical plane we are now – finally – in agreement that equality of opportunity is a desirable goal. At the same time, however, we find ourselves in the grip of a social ethos that insists that not only should we enjoy equal rights but also that we are all virtually identical, differing only in upbringing.

Mercifully, joyously, each of us is a unique individual, and this uniqueness extends to the ethnic and national groups that we form. We are not identical machines with differing software. Without exception, all ethnic groups have produced winners as well as losers in the genetic lottery. Interventionists argue that it is our moral duty to do our utmost to pass on to our children – not the same heritage – but the best, unique heritage possible for each of them. Anti-interventionists point out that, in breaking off the precious baton handed on from generation to generation, we can easily produce an irreparable disaster. But no decision is also a decision.

Many of our everyday decisions are fraught with genetic consequences. Who is having the babies, and how many? Anything that influences fertility is a factor in the new selection. This can include a stroll to the nearest pharmacy to purchase contraceptive devices, a visit to an abortion clinic, or a decision to reduce or even renounce childbearing so as to be able to advance career and education. In denying free day care and financial child support to all but the welfare population, government provides incentives to some groups to bear children and disincentives to others, and this policy has already become a momentous factor in genetic selection.

Eugenicists argue that we must accept our place within the physical world – as biological creatures. To survive as a species with greater philosophical significance than the other animals, they believe we have no choice other than to agree in the area of reproduction to subordinate our interests to those of future generations and begin to manage our populations according to principles that are uncontested when applied to all other species. In short, they advocate replacing natural selection with scientific selection. In the words of Sir Francis Galton, the "father" of eugenics and statistics,

> *What nature does blindly, slowly and ruthlessly, man may do providently, quickly, and kindly. As it lies within his power, so it becomes his duty to work in that direction.*[1]

This book concerns the meaning of life and intelligence and our place in the universe. It is based on a rational philosophy of life and love for our children, of a consciousness of the burdens and responsibilities of parenthood. It is proffered in a spirit of collegial friendship to concerned men and women of good will – both the proponents and the opponents of the eugenics movement. Hopefully, many of them will share the same values, hopes, and fears. If nothing else, we should be able to agree on the right to disagree.

Fraught with history, values, and emotions, the eugenics movement sees itself as based on science, but is not limited to science. I will here attempt to tie together a number of fields in a syncretic approach. I ask the reader's understanding in presenting areas which might seem disparate, but any serious, wide-ranging worldview is necessarily eclectic.

Humankind has entered into the first stages of a revolution in the general understanding of genetic mechanisms, new biotechnologies, and scientific explanation of areas of human health and behavior previously viewed through a moral prism. The genie of enlightenment cannot be squeezed back into the bottle of ignorance. The prospect of holding in one's hands in a few years time the complete human blueprint is awe-inspiring, and we must assume that future discoveries in the field of genetics will give us capabilities that we can barely imagine now. Disagreements on what is attributable to nature and what to nurture will seem quaint, and we will have to ask ourselves as a species what to do next, how to achieve, if not utopia, at least something closer to it than we now have, or at the very least how to survive.

Proponents of eugenics see their cause as part of the struggle for human rights – the rights of people who will come after us. Like Martin Luther King, they argue, we may well wonder whether we will ever reach the Promised Land. Perhaps there is no final goal, just the search, but we owe it to our children to begin the journey, to do our best to ensure that they will be born better people than we are, and that they inherit more of our good qualities and fewer of our flaws.

What Is Eugenics?

This weeping willow!
Why do you not plant a few
For the millions of children not yet born,
As well as for us?
Are they not non-existent, or cells asleep...
Edgar Lee Masters, "Columbus Cheney,"
in "Spring River Anthology"

Once the continuity of humankind with the rest of the animal kingdom was established, invigorated attempts to improve the human genome became inevitable. Eugenics is, after all, quite simply, applied human genetics. Five of the first six presidents of the American Society of Human Genetics were also members of the board of directors of the Eugenics Society. Historically, modern genetics is an offshoot of the eugenics movement, not the reverse.

Positive eugenics refers to approaches intended to raise fertility among the genetically advantaged. These include financial and political stimuli, targeted demographic analyses, *in vitro* fertilization, egg transplants, and cloning. Pronatalist countries (that is, those that wish to stimulate their birth rates) already engage in moderate forms of positive eugenics.

Negative eugenics, which is aimed at lowering fertility among the genetically disadvantaged, is largely encompassed under the rubric of family planning and genetic counseling. This includes abortions and sterilization. To ensure that such services are available to all on a nondiscriminatory basis, it is advocated that, at a minimum, persons with low income receive such services on a free basis.

Genetic engineering, which was unknown to early eugenicists, consists of active intervention in the germ line without necessarily encouraging or discouraging reproduction of advantaged or disadvantaged individuals.

Science

Previous Evolution

The wolf, the snake, the hog, not wanting in me,
the cheating look, the frivolous word,
the adulterous wish, not wanting,
Refusals, hates, postponements, meanness, laziness,
none of these wanting.
Walt Whitman, "Crossing Brooklyn Ferry"

The question of where to draw the line between closely related species and subspecies can be resolved differently by different observers. In the case of modern human populations, where scientists tend to pursue conflicting social-political agendas, demarcation lines are hotly contested.

The system of binomial nomenclature established in the eighteenth century by the Swedish botanist Karl von Linné (Carolus Linnaeus) for mapping the relationships among all living things (at least on our planet) lumps together the totality of modern human populations as *Homo sapiens*. All humans alive today, whether bushmen, Australian aborigines, Japanese, Eskimos, or caucasoids, are thus included in a single species, and any discussion of subspecies is regarded with suspicion and hostility. Issued in response to a statement by the rightist French politician Jean-Marie Le Pen on racial inequality, a 1997 statement signed by a group of prominent biologists denied the very existence of race in human populations. Actually, the denial of race had first been made by the eugenicist Julian Huxley in 1935. Again, the assertion had been triggered by political events – in this case the promulgation of Hitler's anti-Jewish pronouncements.[2] Accordingly we now have a single "modern man," and he comes in different colors. It is true that modern genetic studies have shown remarkable similarity among all humans, but for that humans and chimpanzees share something like 99% of their non-duplicative DNA.

Scientists now generally agree that modern human populations have their origins in Africa, but there is considerable disagreement as to whether current intergroup differences are explained by evolution dating back a million years to *Homo erectus* ("multiregionalism") or whether *Homo sapiens* showed up as a relatively late arrival, roughly 100,000-200,000 years ago, and then proceeded to wipe out competing hominid emigres wherever he came into contact with them ("replacement" theory). The degree to which earlier hominid species interbred remains in the area of speculation, in which the multiregionalists have been accused of making a case for fundamental biological differences that amounts to racism.[3] In the words of the scholar Seymour Itzkoff, we are dealing here with a "will to believe [which] is reminiscent of the seduction of intellectuals with abstract ideological models in politics and social thought."[4]

The family trees of the cheetah and the horse provide useful contrasting models. Genetic studies have demonstrated that today's cheetahs display so little diversity that their ancestors must at one time have come through such a narrow bottleneck that only a few individuals were able to perpetuate the species by inbreeding. Horses, by contrast, display tremendous variance as a result of independent taming and breeding in different parts of the world.

Ultimately, genetics is more like a game of chess, where the development of a position is of strictly historical interest and plays no role in determining the game's outcome, than it is like bridge, where success is determined largely by the player's ability to remember which cards were played earlier. The variability so obvious in human populations, even on an intragroup basis, opens the possibility of intervening in human evolution to guide it and even to search for new horizons, regardless of how present variability came about. Where we came from is a fascinating question, but where we are heading is quite another.

Even the replacement school of thought concedes that the human species developed for *at least* some five to eight thousand generations outside of Africa under radically differing conditions of selection. Such a sequence is sufficient to produce significant differences in the various subpopulations. In addition, still greater diversity would have to be postulated on the basis of the biological diversity that must have been in evidence at the time the various populations left Africa. Since human populations have had a far longer time to evolve in Africa than outside the mother continent, African populations display far greater genetic diversity than do other races, and the tiny populations who wandered out of Africa may well have reflected at least part of this diversity. Moreover, the émigrés may have interbred with other hominid species both in Africa and with those that had arrived still earlier. Animal breeders, by comparison, can achieve significant changes in just a few generations. These factors, combined with the professional specialization of modern society and selective mating, represent the chief sources of intra-species variance.

If *Homo sapiens* has been around for perhaps 150,000 years, our future existence may be considerably more ephemeral. Humanity is thus a colony with a beginning and evidently an end and is viewed here, not just as all people alive at any given moment, but as the totality of future people over the entire lifespan of this community. Eugenicists reason that our moral obligations are to all of them, that we are not only part of the planet's ecology, its custodians as well. As the mythologist Joseph Campbell put it, we are no less than its consciousness.[5]

The renowned geneticist James V. Neel studied the society and genetic makeup of the Yanomama of southern Venezuela and northern Brazil and persuasively argued that the structure of their society was typical of human populations during the period when people still lived exclusively in bands, that is, for all but the last 10,000 years. These were small, isolated populations which practiced polygamy and incest, permitting nature to select among a rich variety of genotypes in widely differing environments. Such conditions were conducive to rapid evolution. Panmixia may still be a long way off, and indeed may never be total, but the ever-increasing outbreeding of human populations is reducing human diversity while at the same time creating large populations that are, perhaps, less prone to sudden, major genetic fluctuations.[6]

History clearly demonstrates that social harmony is especially difficult to achieve in the face of diversity, whether religious, linguistic, or ethnic. The great historical crimes have all been instances of group-on-group violence. And when two or more ethnoses are clearly distinguishable from one another, the situation is fraught with even greater stress. The United States, which renounced the monstrous crime of slavery only to retain blatant discrimination for a century, is now attempting to achieve racial equity, but the fear of racial conflict is and will undoubtedly remain both large and, unfortunately, well founded. At the same time the issue has been blurred, racism being defined as a) group discrimination and hatred and b) discussion of intergroup differences. The two topics are really quite different, albeit not unrelated. Society's elites have decided that studies of intergroup differences are too volatile to permit them to be widely discussed and have falsely presented such studies as claiming total separation of group qualities rather than relative statistical frequency of specific characteristics.

We should all be able to agree that intergroup differences are a scientific, not a moral question. As far as the eugenics argument is concerned, they are irrelevant in the most fundamental fashion. Even if the desired breeding resource proves to be distributed differently in some populations than in others, each group contains a vast pool of talented individuals to draw upon in parenting future generations. Regardless of the magnitude of such intergroup differences, the reality is that even on an intragroup basis we ought to be less than pleased with ourselves.

Testing

A sure test, an easy test:
Those that drink beer are the best,
Brown beer, strongly...
Robert Graves, "Strong Beer"

Since IQ testing was first initiated in the early part of the twentieth century, it has been utilized intensively by the US army both to select recruits and to determine the areas in which they might best be employed. Proponents of the egalitarian grain have delighted in attacking century-old science and then applying their conclusions wholesale to modern science. Certainly early IQ tests contained questions that elicit embarrassed smiles among today's testers. For example, was the Knight engine used in the Packard, the Lozier, the Stearns, or the Pierce Arrow? Or does Velvet Joe appear in advertisements of tooth powder, dry goods, tobacco, or soap?[7] While such questions might have had some limited validity when addressed to young people who had grown up in America, they were obviously inappropriate for people who had recently immigrated to the United States and barely spoke English. Such persons performed badly on the test, but it does not automatically follow that modern tests, which have been worked on assiduously by thousands of psychologists, are equally flawed and thus totally invalid.

Hopefully, the massive expansion of education throughout the world in the twentieth century has helped people not only to acquire specific facts, but also to use their minds more efficiently. But the fear is that dysgenic fertility patterns inherent in modern society have created a population with less innate ability than that of its predecessors.

To approach this question we must first make clear the difference between genotype and phenotype. Genotype is genetic potential; phenotype is realized potential. For example, statistics show a constantly rising mean height in most of the world. The cause is obviously not altered genes but improved nutrition (and, perhaps, meat laced with hormones). But genotypes set limits. If a group of pigmies were to be given excellent food and a group of Massai tribesmen were to be distributed low-quality nourishment, the pigmies would obviously enjoy a height increase and the Massai a decrease, but the pigmies would not become taller than the Massai, and there would be no Lamarckian carry-over to their children.

As the psychologist Edwin Boring once quipped in a debate with the columnist Walter Lippman, "IQ is what IQ tests measure." This is not necessarily the same thing as raw intelligence. One must distinguish between a conceptual variable and its operational definition. IQ is simply one possible measure of phenotype.

Some estimates of genotypic IQ decline are in the range of 1 to 4 points per generation,[8] but the New Zealand political scientist James R. Flynn has produced a seminal study claiming that IQ scores have actually been steadily increasing. Such tests as the Stanford-Binet and the Wechsler regularly measure subjects and establish new mean scores and standard deviations. From 1932 to 1978 testers steadily reset norms, each time raising the bar. When the norms are held constant, the mean IQ has risen 13.8 points – nearly one standard deviation over the course of 46 years.[9]

This is a potentially very encouraging result. It indicates that IQ differences may prove to be relatively more malleable than was previously thought, and the egalitarian ideal, which lies at the heart of the eugenic cause, may be more easily realizable than previously believed. On the other hand, we still can only surmise the constraints laid upon phenotype by genotype. What evidently has happened, if Flynn is correct, is a phenotypic improvement that has overridden genotypic deterioration.

The SAT I is intended as an aptitude test, as opposed to the SAT II, which measures knowledge in specific subjects. The SAT I consists of two parts, the SAT V (verbal) and the SAT M (math). Flynn goes on to point out that, simultaneous with the above-mentioned IQ gains, an opposite trend was noted in SAT verbal scores.

SAT scores can be raised by coaching, but improvements are subject to a law of diminishing returns. Math scores rise by roughly 30 points after 40 hours of coaching, and verbal by about 20. But continued improvement of even 50% in scores is not achieved by putting in even six times that number of hours.[10]

Testing has generally enjoyed broad public support. In 1979, the Gallup Organization asked a representative sample of Americans what they thought of standardized tests. Eighty-one percent responded that they were "very useful" or "somewhat useful."[11] At the same time, a powerful coalition of the National Education Association, National Association for Colored People, and Ralph Nader's followers adamantly opposed them. The coalition had many influential supporters in government and the press. Dan Rather, for example, in the 1975 CBS news special *The IQ Myth* declared that not only were IQ tests relatively useless as measures of intelligence, but that they were biased as well, for "it's economic class that marks the main dividing line on IQ scores."[12] But the coalition did not have the general support of one group that is allied with it on many other issues. Jews invariably come off well in testing, and thus it is not surprising that the American Jewish Committee, the Anti-Defamation League, and the American Jewish Congress have all filed amicus briefs with the Supreme Court in opposition to Affirmative Action programs.[13]

g-loading

Lord, make me to know mine end,
and the measure of my days, what it is;
that I may know how frail I am.
Psalm XXXIV, 4

Does such a thing as *general intelligence* ("g") exist, or does each individual possess a disparate collection of unrelated abilities – that is, multiple intelligences? Any scientific discussion of "unitary intelligence" is fraught with political significance for it can be interpreted as providing the measure of a person's overall worth or ranking.

Proponents of general intelligence, beginning with Charles Spearman in the early twentieth century, have pointed out the positive correlation between spatial, numerical, and verbal abilities. An IQ score is essentially a numerical expression of g. On the other hand, there is no denying the existence of *idiot-savants* – people who have difficulty in coping with even the most elementary everyday tasks but who may be accomplished musicians or sculptors, can add a series of numbers with no less precision than a calculator, or can easily recount weather conditions on a randomly selected day in the eighteenth century. In other words, the correlation between their one special ability and their other abilities is negative. And we need not limit ourselves to the exceptional. When specialized aptitude tests were administered to a group of students in place of global measures of intelligence, more than half of them scored in the top 10% on a specific ability.[14]

How then to compare or evaluate disparate abilities? The significance of g-loadings may well be exaggerated – or even a *non sequitur*. Given the limited physical space occupied by the brain, hyperdevelopment of certain abilities may even necessarily come at the expense of others. In many ways the question is like the proverbial glass which is either half-empty or half-full. It all depends on the observer's point of view.

IQ Decline

Tis folly to decline,
And steal inglorious to the silent grave...
Sir William Jones,
"An Ode: In Imitation of Alcaeus"

How can we best protect the interests of still unborn generations? This is extremely difficult in a world where many regard children as an ordinary commodity. The so-called "demographic transition," in which people in advanced societies choose to have fewer children, is even studied by economists and demographers in all manner of curves, graphs, and charts, establishing the cost of one child as the equivalent of X number of automobiles, televisions, or what have you.

What are the consequences for the gene pool of selecting out young women of ability to pursue education and careers, thus reducing their fertility (in 20% of U.S. couples, delayed fertility turns out to be cancelled fertility) while remunerating young women of lesser ability on the basis of how many children they bear, even denying them abortions when they themselves request them?

Whereas girls in countries with developed welfare programs can choose to escape school by becoming pregnant if they find themselves unable to cope with an academic program, an early 2001 study showed that fully a third of American women earning more than $55,000 a year are childless at age 40 and are likely to live out their lives without ever giving birth.[15]

While "Total Fertility Rates" (TFR – the number of children a woman has in her lifetime) represent an important yardstick in measuring fertility patterns, generational length also plays a role. Obviously, the earlier a woman begins having children, the more offspring she can bear. Imagine two groups, in one of which women have their children at the average age of 20 and the other at 30. The first group will effectively have 50% more children than the first even if the TFR is identical. In the New York Longitudinal Study of Youth, for example, women in the bottom 5% of intelligence had their first baby more than seven years earlier than women in the top 5%.[16]

Abortion is significant in terms of the eugenics argument to the degree that it affects selection, particularly when the service is readily available to high-IQ groups, who can easily pay for it, but is denied to low-IQ groups, who are dependent on receiving the service on a subsidized or free basis. The abortion rate is related to years of education, which can be used as an imperfect substitute for IQ. In 1979, the standardized U.S. abortion rate by years of education for women 20 years of age and older was 44.3 for women with a high school education but only 3.2 for those who had less than eight years of schooling.[17]

Another significant dysgenic factor is war. The creature who sees himself as molded in the image of God has used his improved technology to do vastly greater violence not only to his environment but also to himself. And it has been the egalitarians, not the hereditarians, who have been the least squeamish about murder and exile, be it in Russia, China, or Cambodia. There is a sad consistency to their logic: if everyone is the same, anyone who interferes with achieving utopia in our time can simply be eliminated and replaced when the next generation shows up.

War as a destructive mechanism of natural selection became a frequently discussed topic when "the flower" of Europe's youth marched off to die *en masse* in the trenches of World War I. It was, after all, this particular conflict which introduced IQ testing to select out young men of ability more accurately for use as cannon fodder.

In instances of violent civil conflict, too, force is targeted most heavily at the real and potential opposition. Since opposition by definition involves thought and ideological dedication, the targets of destruction, more frequently than not, are persons of ability. The historian Nathaniel Weyl christened the phenomenon "aristocide."[18] Statistical analysis demonstrates that such a process produces a relatively modest lowering of the mean population IQ, but disastrous reductions in the number of persons with exceptionally high scores.[19]

The contribution of outstanding individuals to culture, science, and the general quality of life is disproportionate to their numbers. Just imagine what the history of music would be like without just a handful of the great composers – Bach, Beethoven, Mozart, Brahms, Stravinsky, Mendelssohn. The same sort of "short list" could be made up of physicists, mathematicians, philosophers. Eliminate these geniuses and the average ability level of the next generations will not be altered perceptibly, but how impoverished our world would be!

The consequences of such a process are obviously alarming. Even with a relatively stable mean IQ, a society in which the intellectual leadership is significantly reduced is an impoverished society – at least relative to its original state. The lesson to be drawn is that the turbulence and magnitude of social upheaval do not have a necessary relationship to their genetic consequences.

Genetic Illnesses

There is no such thing as immutability in biological stocks, for with each new generation a species inherits new genes in the form of mutations. On rare occasions a mutation can improve the individual's survivability chances, and the new gene then becomes more widespread in the population as a whole. Nevertheless, the vast majority of mutations end up reducing the number of offspring. This is the classic balance of mutation and death which is called "natural selection," and it is accepted by biologists as decisive in all species.

This book aims to pose certain broad philosophical questions about the values and goals of human civilization and the path which humankind will follow in consciously choosing either to pursue or to reject artificial selection. It is not intended as a discussion of the complexities of human genetic disease. By way of analogy, one could compare this document to a roadmap rather than to an automobile repair manual, but a few particularly important nuts and bolts still need to be mentioned.

We have made such advances in medicine that natural selection has been reduced to almost zero. Already 98% of Americans survive at least to their twenty-fifth birthday.[20] Medicine is intended largely to benefit its creators – the currently living. Thus, if we speak about illness, the emphasis is on "horizontally transmitted" infectious diseases over "vertically transmitted" genetic diseases. It is, after all, very difficult for a doctor, a pharmaceutical company, or a hospital to collect a fee from people who have yet to be born. Medicine is a business that depends on paying clients, and the most motivated clients – those who not only can but who are eager to pay – are the ones who are hurting now.

The *Encyclopedia Britannica* succinctly presents some of the salient facts related to the 3,500 autosomal dominant, autosomal recessive, and sex-linked disorders that have already been catalogued (the list is rapidly expanding):

> *Epidemiological surveys suggest that approximately 1 percent of all newborns have a single gene defect and that 0.5 percent have gross chromosomal anomalies severe enough to produce serious physical defects and mental retardation. Of the 3 to 4 percent of newborns with birth defects, surveys indicate that at least half suffer a major genetic contribution. A minimum of 5 percent of all conceptions that evidence themselves have gross chromosomal anomalies, and 40 to 50 percent of spontaneous abortions involve chromosomally abnormal embryos. About 40 percent of all infant mortality is due to genetic disease; 30 percent of pediatric and 10 percent of adult patients require hospital admission because of genetic disorders. Medical investigators estimate that genetic defects – albeit often minor – are present in 10 percent of all adults.... About 20 percent of all stillbirths and infant deaths are associated with severe anomalies, and about 7 percent of all births show some mental or physical defect.*[21]

It gets scarier. Spontaneous mutation rates, genetic "typos," have been estimated at 200 per person,[22] most of which appear to be neutral, but an unknown percentage of which are undesirable when expressed, their effects being cumulative. Aside from genetic anomalies which are necessary and sufficient to cause a specific illness, a much larger number of multifactoral illnesses exist in which certain genes create a disposition toward specific illnesses, for example, most cancers, diabetes, and hypertension.

Early eugenicists had the naïve notion that simply to prevent persons suffering from genetic illness from having children was sufficient to produce a healthier population with each generation; however, most genes which cause diseases are both recessive and extremely rare. Thus, the number of carriers greatly outnumbers the number of persons actually affected, and the nonreproduction of actively ill individuals could achieve only an extremely slow reduction of the disease in subsequent generations. This means that if an undesirable trait occurred in 1% of the population it would take 90 generations to reduce the incidence to 0.01 and 900 generations under conditions of random mating to achieve a reduction to the level of one in a million.[23] Even then, a natural spontaneous mutation rate would remain, which would also have to be countered on a never-ending basis.

Genetic engineering techniques are advancing rapidly. It is already possible for carriers of genetic diseases to conceive children *in vitro*, then perform embryo screening, known as preimplantation genetic diagnosis, and select a healthy embryo for implantation in the mother's womb. This is a eugenic technique which is already being implemented on a voluntary, gradual basis. In the not so distant future it will be possible to make changes in the germ cells (those involved in reproduction), and not just in the somatic cells (those not involved in reproduction). Germ-line therapy does not fit into either positive or negative eugenics, both of which amount to encouraging or discouraging an individual from entering into the sequence of generations, but such therapy is unquestionably eugenics. When the possibility first arose, the general attitude was one of absolute condemnation; now the tendency is to speak more in terms of a moratorium of this new therapy. The bioethicist Fritz Mann at the Free University of Brussels writes:

Aside from religious grounds, there exists no ethical justification for not influencing the germ line. If one day a cure is discovered for healing a hereditary disease in this fashion, not only for its bearer, but for all his descendants, what reason could there be for forbidding it?[24]

Such an achievement will represent a genetic breakthrough, but the puzzle of genes and their interactions is only beginning to be solved. Nevertheless, geneticists are already altering the germ lines of plants and animals, and human germ-line therapy is only a question of time. Meanwhile, genetic counseling and treatment are on occasion helping those alive today at the expense of future generations. A prospective parent who knows that he or she is the carrier of a recessive gene which can cause illness in subsequent generations, can selectively abort fetuses in which the gene will be actively expressed. Thus, the immediate children of the union are free from the illness, but the number of carriers of the recessive gene increases further down the generational chain.

The question is whether parents have a moral right to bring children into the world who will be disadvantaged by their heredity. To quote the philosopher Emmanuel Lévinas, "my son is not simply my creation, like a poem or an object. He is not my property."[25] Can parental responsibility be sloughed off, denied? Marcus Pembrey, a professor at the Institute of Child Health at the University of London, in discussing genetic counseling argues that

The aim should not be to reduce the birth incidence of genetic diseases, because to make that the objective of the services would be to by-pass the mother's choice in the matter of selective abortion... The view that reduction in the birth incidence of genetic disorders is not an appropriate objective for genetic services is finding wide acceptance.[26]

This is the so-called "personal service model"[27] of genetic counseling, which subordinates children's well-being to that of their parents. Such a view could well be challenged in the courts, perhaps in *wrongful life* legal suits (which first appeared in the United States in 1964, claiming *wrongful death* suits as a legal precedent) or even on a class-action basis. Whereas we may have previously lacked the knowledge to reduce genetic illnesses, the ignorance argument will have less and less weight in the future. The parental appeasement posture will not be comparable to the Thalidomide baby scandal of 1957-1961, for this will be an act committed with full knowledge and intent.

Germ-line interventions will encounter resistance from people who feel, some on religious grounds, that such therapy is "unnatural" and that we have no right to "play God." Even conventional care is rejected, for example, by certain religious groups, and one occasionally comes across newspaper articles describing a family whose child has died for lack of medical treatment. There will also be nonreligious objections by people who are wary of making mistakes. Indeed errors are a real possibility. When we will have achieved a much better understanding of human genetics, however, the nonreligious objectors will have considerably less wind in their sails.

Israel has been a forerunner in genetic counseling. In the words of a researcher at Ben-Gurion University, "Eugenic thinking is alive and well [in Israel] today."[28] Gideon Bach, head of Genetics at the Hadassah-Hebrew University Medical Center in Jerusalem commented:

> *We now know that most, if not all, human disorders have a genetic background, and we're acquiring the tools to study, treat and eventually prevent or cure them.... Israel, with many inbred ethnic groups, has proven a rich human laboratory for genetic detectives. It's far easier to trace genetic anomalies in inbred groups with homogeneous pedigrees.*[29]

Ashkenazim, who until some forty years ago largely intermarried, carry a dozen recessive genetic diseases with relatively high frequency. The best known is an autosomal disorder christened Tay-Sachs after its description in 1881 by the British ophthalmologist Warren Tay. It is caused by the hereditary lack of a crucial enzyme that normally breaks down fatty waste products found in the brain. If both parents are carriers of the gene, the child has a 25% chance of suffering from the disease, and a 50% chance of being a carrier. One in 27 Jews in the United States carries the gene. A baby suffering from the disease at first appears normal, but becomes hypersensitive to sound after a few months. Eventually the child becomes deaf, blind, mentally retarded, and unresponsive to outside stimuli. Death results by age five.

In 1985, Rabbi Joseph Eckstein, citing the Bible and the Talmud, founded the international genetic testing program call Dor yeshorim ("generation of the righteous") with the goal of preventing further children from being born with the illness. In the program, Orthodox Jewish students are tested to determine if they carry the gene. If only one prospective parent is a carrier they are not advised against marriage, but if both test positive they are counseled to choose a different marriage partner.

Israel has one of the highest screening rates in the world, testing well over ten thousand people a year.[30] The writer Naomi Stone expresses what is evidently the general Jewish attitude toward prevention of Tay-Sachs:

> *Perhaps, the disease can be eradicated entirely from populations where it is concentrated, and if this were the case, who could reasonably express qualms?... I am an Ashkenazi Jew, and I know that it is my obligation to be acutely aware of my heightened risk factor for the disease.*[31]

Understandably, eugenic practices in the United States are often resisted among representatives of the handicapped community. Bioethicist Adrienne Asch writes:

> *My moral opposition to prenatal testing and selective abortion flows from the conviction that life with disability is worthwhile and the belief that a just society must appreciate and nurture the lives of all people, whatever the endowments they receive in the natural lottery.*[32]

Much the same position is held by the Canadian ethicist Tom Koch, who believes that all diseases are part of the diversity of the human race.[33]

Gregor Wolbring, another Canadian active in the movement of handicapped persons against eugenics, goes even further:

I can say, without hesitation, that my life has been richer because I have MS. How can anyone who has no experience with disabilities understand that?[34]

Mr. Wolbring, who runs a website with materials both supporting and attacking the eugenics movement[35], points out that he himself is opposed to eugenics.

Another internet document reads:

The underlying issue in eugenics is that someone decides, based on stated or unstated values, which characteristics are worthy enough to be part of society and which are not [Discrimination]... The key question is how a society (social eugenics) or a person (personal eugenics) decides which characteristics are permissible in an offspring/offspring to be. Can a society influence or regulate the decisions of social/personal eugenics? Is there a rational way to distinguish between Tay-Sachs, beta-Thalassemia, sickle cell anemia, thalidomide, Alzheimer, PKU, gender, sexual orientation (if a way were ever found to predict it), mental illness, cystic fibrosis, cerebral palsy, spina bifida, achondroplasia (dwarfism), hemophilia, Down Syndrome, coronary heart disease, osteoporosis, and obesity?... A war of characteristics is on, which will disenfranchise many characteristics from the human rights movement and from equality rights. This has to stop."[36]

While this anonymous author does indeed raise thorny questions with regard to certain characteristics – for example, sexual orientation, dwarfism, and obesity – the defense of some of the named horrendous diseases is disconcerting, albeit stemming from a legitimate and well-founded fear of discrimination against the persons who suffer from them. It is our duty to ensure that we indeed discriminate against the disease and not against the victims.

Scientific Method

Any attempt to channel the sexual act requires that society first dismantle the devilish scaffolding of taboos, phobias, neuroses, and fetishes that has been erected around human reproduction.[37] Given the fundamental continuity of the human animal with the entire biological kingdom in general and with mammals specifically – including such intimately related species as the higher primates – the revolution in developmental and molecular biology is resetting the intellectual climate by conceptualizing human reproduction in accordance with the principles of animal breeding.

Genetic selection presupposes genetic variation; otherwise there would be nothing to select from. Heritability is the yardstick by which both natural and artificial selection are measured. Heritability scores are mathematical correlations ranging from 1 (a parental trait is inevitably passed on to the children) to 0 (the children are no more or less likely to possess it).

The heritability of economic traits has been intensively studied for farm animals. For example, milk production is 0.25, yearling body weight in sheep is in the range of 0.2 - 0.59, and feedlot gain in beef cattle is 0.5 - 0.55.[38] The heritability for height among white European and North American populations is 0.9.[39] Using data from twin studies, Thomas Bouchard and colleagues at the University of Minnesota have placed the overall heritability of personality at about 0.5. Heritabilities of social attitudes are even higher: 0.65 for radicalism, 0.54 for tough-mindedness, and 0.59 for religious leisure time interests. Occupational interests correlate at about 0.36.[40] One study of monozygotic (identical) and dizygotic (fraternal) twins showed that monozygotic twins showed a significantly higher correlation than dizygotic twins for being frank, active, talkative, gregarious, extroverted, assertive, calm, self-confident, even-tempered, emotionally stable, kind, polite, pleasant, agreeable, thorough, neat, systematic, conscientious, inventive, imaginative, original creative, open to experience, refined, sophisticated, and flexible. Model-fit analyses suggested about 40% genetic, 25% shared environmental, and 35% nonshared environmental influence.[41]

Although the heritability of any trait or combination of traits can be measured along this same scale, it is the intelligence controversy which has attracted the most heated attention. Low estimates of IQ heritability in human populations are generally on the order of 0.4, with 0.8 being the ceiling for high estimates.

How to disentangle nature from nurture? The correlation between the IQ scores of the same person taking the same test a second time can serve as a benchmark; it is 0.86.[42] The prominent English psychologist Cyril Burt located a number of identical twins who had been raised separately. In 1966 he reported an IQ correlation of 0.77 among 53 pairs of identical twins whom he had studied. When Burt, who died in 1971, was posthumously accused of having falsified his data, the purported scandal made for major news. Now, however, a great deal more research has been done on the topic, and Burt's findings have been replicated repeatedly, including Bouchard's study of 8,000 twin pairs, which came up with a correlation of 0.76 for identical twins reared separately and 0.87 for those reared together. [43] In another study of adopted children, conducted by Sandra Scarr and Richard A. Weinberg, also at the University of Minnesota, the adoptees' IQ scores correlated significantly more positively with those of their biological than with those of their adoptive parents.[44]

Natural selection depends not only on genetic variation but also on environmental variation. The greater the range of the two forms of variation, the greater the intensity of selection – that is, the faster the rate of evolution. For millennia now, without any knowledge of Darwin's theory of evolution, people have been able to pursue artificial selection successfully in plants and animals by simply breeding the most desirable individuals with each other under the principle "like breeds like." This is still the chief methodology of animal breeders. When, however, low variation or low heritability impede selection, modern genetic tools are employed: frozen semen, separation of male- and female-producing sperm, superovulation, embryo storage and transfer, *in vitro* fertilization, and transfer of genetic material.

The use of artificial insemination renders eugenic measures applied to males far more effective than to females. For example, by employing modern techniques a bull can theoretically be made to produce 200,000 breeding units of semen per year.[45] One bull already has 2.3 million granddaughters.[46] Furthermore, sperm can be frozen for long-term storage and later use.

If there is no shortage of premium-quality sperm, the same is also true of eggs. Only a tiny percentage of the eggs created in human females at birth are ever fertilized. *In vitro* fertilization, with resulting embryos implanted in a womb other than that of the original mother, would make it possible to achieve a revolution in population quality without creating a quantitative bottleneck.

Cloning is a still newer technique. During the process a genetically identical copy of a biological organism is produced by asexual means. Cloning is common in nature. Any plant that can grow from a cutting, or animal tissue that can reproduce itself in a Petri dish, in the process also produce clones.

During laboratory cloning ("nuclear transfer"), the genetic code of an individual organism is inserted into an egg that has been stripped of its own nucleus, and that egg is then implanted in the womb of a "birth mother," just as is already done in cases of *in vitro* fertilization. The child who is born is the donor's identical twin. The first animal clones were produced in the late 1950s. In 1993 US researchers experimentally cloned a human being as a possible treatment for infertility, but the experiment raised a storm of criticism. The cloning of the sheep "Dolly" did not take place until 1996. Other mammals already cloned by scientists include horses, dogs, rabbits, cows, goats, deer, pigs, cats, rats, and mice.

The current debate on cloning is focused on therapeutic cloning. For example, it may be possible in the future to clone cells from a person suffering from cardiac insufficiency, develop those replacement cells into heart muscle, and then transplant that muscle back into the same patient without fear of rejection.

The real issue, however, is reproductive cloning – conceiving babies who will be brought to term and who will enter the general population as independent persons. Reproductive cloning can be pursued for two reasons: first, as a device to combat infertility, and second, to enrich the human gene pool. I refer here to the latter as "eugenic cloning." Cloned embryos, as well as embryos produced during in vitro fertilization, could be implanted in a womb which might be human, animal, or even artificial. "We can see all too clearly where the train is headed, and we do not like the destination," wrote Leon Kass, chief of George W. Bush's Bioethics Council.[47] Revealingly, Kass, who is an observant conservative Jew, has also come out against the dissection of cadavers, organ transplantation, in-vitro fertilization, cosmetic surgery, and sexual liberation. Virginia Postrel, editor-at-large of *Reason* magazine, responded to the views expressed by Kass by commenting that "This isn't about the 20th century. It's about the 16th."[48]

Much of the criticism of cloning stems from a fundamental misunderstanding – that there is an intent to produce a race of identical creatures lacking any and all individuality. This is definitely not the case, and no such practice has ever been advocated. Rather, it is the expectation that persons born as the result of a cloning process would enter into normal sexual relations with the vastly greater population of individuals born as the result of traditional sex and would multiply in the traditional fashion, thus increasing the frequency of advantageous genes in the following generations.

Despite some well-publicized successes, there remain a number of difficulties to be worked out, and the failure rate is still high. For example, cloned animals often have abnormal placentas – a factor that affects size and survival. Part of the problem evidently lies in abnormalities in gene expression.

Much of the resistance to cloning comes from religious groups, but is not limited to them. Aside from a fully legitimate fear that we may still not be knowledgeable enough to proceed immediately to human cloning, the resistance to cloning per se is startlingly reminiscent of the traditional argument against evolution – that it is "an assault on human dignity." That was precisely the text and heading of an open letter addressed to President George W. Bush in the *Washington Times* in January, 2002, signed by 29 conservative political and religious leaders.[49]

The media have waged an energetic campaign against cloning. We have examples in the 1976 novel, The Boys from Brazil by Ira Levin, made into a film starring James Mason in 1978, and most recently in 2002, with the appearance of *Star Wars Part II: Attack of the Clones*. There is even a canard as to whether human cloning methods might be patentable.

The *New York Times* is entirely correct: "Opposition to reproductive cloning is universal in Congress,"[50] and if any senator or congressman secretly harbors a more benign view of the procedure, the chance that he or she will express that opinion publicly is absolutely zero. In 2001, the House of Representatives voted to ban all forms of cloning, but the Senate resisted a total disallowment. Congress has thus resolved to criminalize reproductive cloning, even though Congress's unanimity in this area is not shared by everyone in the scientific and scholarly community. According to the *Wall Street Journal*, "some diplomats said they believe the U.S. stand in the U.N. was primarily intended to score domestic political points with religious conservatives and antiabortion activists."[51] But such moods are hardly limited to the United States. On November 6, 2003, by a 80-79 vote, with 15 abstentions, the United Nations narrowly resolved to delay by two years a vote supported by the United States and the Vatican to outlaw both therapeutic and reproductive cloning. A number of other countries supported a Belgian proposal to ban reproductive cloning while permitting therapeutic cloning.

Animal breeding methods usually amount to producing a specific type on the basis of very strict characteristics. The same is true for plant selection, in which a rich variety of strains is usually replaced by a few monocultures. Nothing of the sort would be appropriate for human populations. Human selection, as proposed by proponents of eugenics, would be aimed at a far more limited reduction in genetic variance. Diversity is viewed not simply as a great source of strength but also as an integral part of what we are and want to be. A certain reduction of this variability, on the other hand, is the mathematical goal. Eugenicists argue that even a very significant channeling of motherhood and a far more stringent selection among men would still leave billions of people reproducing. By comparison, all thoroughbred race horses stem from three Middle Eastern stallions, and natural selection can be even more draconian.

Mapping the Human Genome

We have the intestines of chickens
to tell the fortunes of war.
We have slaves
that they might be silent.
We have stones
that we might build.
Why then should we trouble the gods?
Osip Mandelstam, "Nature is the Same Rome..."

Genetics is a very young science. The theory of evolution was not forwarded until the late 1850s. In 1866 the Austrian monk Gregor Mendel had begun to attempt to pry open the secret of creation when he published the results of his controlled pollination of the garden pea, but his discoveries were ignored for the rest of the century, and Galton never learned of them. Even the discovery of the mechanism of fertilization as a union of the nuclei of male and female sex cells was not made until 1875; 1888 saw the discovery of certain deeply stained bodies in cell nuclei, which were christened "chromosomes," and in 1909 the word "gene" came to be applied to the Mendelian factors of heredity. The first *in vitro* fertilization (rabbit and also monkey) was not achieved until 1934, and as for the double helical structure of DNA, its discovery dates back only to 1953. This is all so recent that although early eugenicists had set their goals and methods they were largely ignorant of the mechanisms involved.

The mapping of the human genome is still in an early stage. The amount we don't know vastly dwarfs what we do know. There appear to be approximately three billion bases, or chemical letters, making up the nucleotide sequences that form 20,000 to 25,000 genes which code directly for proteins. Just how genes and the proteins they produce interact is still poorly understood.[52]

But protein-coding genes comprise only 2% of the human genome. The functions of other DNA sequences are still largely a mystery. We do know that some of them contain switches that turn genes on and off, and we have learned that at the ends of the chromosomes there are telomeres, whose shortening appears to be related to the aging process, and nonfunctional genomic parasites, whose only function in our bodies seems to be to replicate themselves. An estimated 40-48% consists of repeat sequences. Even after sequencing the genome, we will still have to determine how these data relate to expression. The sequences are only the parts list to a grand machine, the outlines of which we are only beginning to trace.

Scholarly opinion is rapidly growing more cognizant of the role of genes in human society. In 1998, University of Massachusetts political scientist Diane Paul wrote that just fourteen years earlier, in 1984, she had labeled as

> *"hereditarian" or "biological determinist" the view that differences in mentality and temperament were substantially influenced by genes – employing these terms as though their meanings were unproblematic. That usage today would surely be contested. For the view implicitly disparaged by these labels is once again widely accepted by scientists and the public alike.*[53]

The bottom line is that with every day we gain greater knowledge and that in the not all that distant future we will be able to predict, with a high degree of certainty, the *genetic load* that we are passing on to future generations.

Ideology

Essential Conditions

For we know in part, and we prophesy in part.
I Corinthians, xiii, 9

Proponents of eugenics see the movement as an integral component of an environmentalist policy. They reason that, while we cannot predict the distant future, we can with a fair degree of confidence trace out certain conditions which will always be essential or at the very least desirable:

- a supply of natural resources,
- a clean, biodiverse environment,
- a human population no larger than the planet can comfortably sustain on an indefinite basis,
- a population which is healthy, altruistic, and intelligent.

The blessings that we are reaping from the industrial revolution are, to a significant degree, unsustainable. We are systematically depleting the planet's riches. Debates as to how long this or that resource will hold out are essentially trivial in the greater scheme of things, for eventually we will have thoroughly sifted through the earth's accessible subsoil. The only resources that we can count on over the long run are those which are truly renewable or inexhaustible. As for science-fiction fantasies about relocating to other planets, this "trash-the-world" vandalism is unfeasible for billions of people.

Of course, it can be argued that the inevitability of resource exhaustion makes it a non-topic. What is the difference if this process is completed sooner or later? The eugenicists' response is a moral one. We embarked upon the industrial revolution only two centuries ago, and we have a huge transition to go through if we do not wish our offspring to return to a hunter-gatherer economy in which there will be precious little left either to hunt or to gather. We need to husband our precious, finite resources to get through this transition in as chary a fashion as possible.

Traditional societies live in harmony with nature. Modern industrial society clearly does not, and we have already overwhelmed much of Nature's ability to heal itself. An enormous number of species have been wiped out, while still others have been transported by man to different environments where, lacking natural enemies, they have followed the example of man in replicating his devastation. Globalization is already delivering devastating blows to the planet's biodiversity. As for pollution, while it has gone so far that it becomes too painful to even read about in the papers, much of it can still be reversed.

And there are population problems which may overwhelm the planet in a relatively short period. In traditional societies children, being the only form of social security around, represent for their parents an economic good. More is better. In economically developed societies, on the other hand, children are strictly an economic liability, and the surest way to maximize consumption (for many the ultimate purpose of life) is at the very least to reduce the number of children.

In 2003, the Total Fertility Rate in East Asia was below replacement at 1.7. The national TFR had even dropped to 1.3 in Japan and Taiwan. Europe's TFR had fallen to 1.4. Canada's and the United States' TFR were 1.5 and 2, respectively. In sharp contrast, Latin America's TFR was 2.7, while Africa's was 5.2. The global TFR was 2.8, the planet's population having swollen six-fold over the last 250 years. It is still growing by leaps and bounds, although more slowly than formerly. The largest growth is taking place in the poorest countries. While it is hoped that the entire world will eventually pass through the demographic transition, it is not impossible that before this happens individual countries will undergo horrendous Malthusian collapse. Bangladesh, for example, which has a population of 134 million on a land mass roughly the size of the state of Wisconsin, most of which is an alluvial flood plain frequently ravaged by hurricanes, is projected to increase its population to 255 million by the year 2050. Other countries provide even more rapid growth rates: The Palestinians during the same period are predicted to increase their numbers to form a population 3.3 times its current size, and this on land where water is already in critical shortage. India is projected to add as many people as Europe's entire population by that time.[54]

Demographic predictions are not made with any claim to precision. There are low, medium, and high projections. And there are questions to which no one has any answers. What is the long-term carrying capacity of the planet? How many lives will be carried off by phenomena that reduce the population not by decreasing fertility but by increasing mortality? Already there are projections of a loss of fifty million deaths from AIDS. Where will it end? What new plagues lurk around the corner? Military conflicts could easily result in the deaths of billions of people. Demographic predictions are really no better than stock market predictions. In any case, eugenicists argue that the wisest approach is to err on the side of caution. A smaller population capable of surviving by the use of current renewable resources will create less stress and make the transition to a new economy more manageable.

Altruism

You among the dry, dead beech-leaves, in the fire of night,
Burnt like a sacrifice, you invisible...
D. H. Lawrence, "Scent of Irises," 1916.

Darwin pointed out that natural selection favors behavioral patterns which promote survivability. Suicidal behavior, it would seem, should lead to the destruction of the animal involved, thus preventing it from reproducing. How then, sociobiologists asked, could the behavior of a honeybee be explained when, in stinging a perceived threat to the hive, it rips out its own belly together with the stinger and thus perishes? The answer is that survivability of the genotype, not of the individual, is crucial. Although the individual bee dies, the other members of the hive are genetically identical copies, and the chances for the survival of their genes are improved by the sacrifice of the individual.

Up until recently, survival of a human individual was problematic. People are physically unimpressive animals, with easily torn skin, no claws, weak musculature, and atrophied canines. In primitive times opportunistic out-of-clan cannibalism would have improved survival chances. Thus, such individuals or groups would have been viewed not merely as enemies but as potential food. We are the products of precisely such an evolutionary process.

In all animal species, out-of-family altruism is the rare exception. Survival requires maximum expenditure of effort, and efforts expended on alien genes (dispersed or nonfocused altruism) waste effort and thus, by definition, reduce survivability.

Most traits are arranged along a continuum, and altruism is no exception. If a statistical curve were drawn to display diffuse altruism at one end and focused altruism at the other, the result would be radically skewed toward focused altruism – that is, toward immediate offspring.

As man moved into larger groups (tribes), specialization and cooperation went hand in hand. The skew was retained but became less pronounced, and people learned to "live by the rules" and even to feign nonfocused altruism. But the genes really didn't really change all that much. *Homo sapiens*'s political history presents an unbroken string of violence, and any objective determination of his coordinates within the animal kingdom places him among the predators.

What sort of a society do we want? To the degree that altruism is determined by our genes, artificial selection could theoretically make it possible to create a social profile skewed toward diffuse altruism. The difficulty of working toward a better society is that such a process necessarily entails effort and even sacrifice on the part of the currently living, who have the power of absolute dictators.

All this leads to gloomy conclusions. Professor of human ecology Garrett Hardin wrote that it is futile to expect people to act against their own self-interest,[55] and the bioethicist Peter Singer defines "reciprocal altruism" as merely a "technical term for cooperation."[56]

The big question, of course, is how to select for altruism. The same questions must be answered here as for other traits. How to measure? What are the relative contributions of nature and nurture? Which genes come into play and in which combinations? What is the heritability? What combinations of positive and negative eugenic approaches are likely to prove most effective?

A good Trekkie, the eugenicist wishes to create a global civilization which does not set consumption as its primary goal but longs for a loving, nonpredatory society that pursues the goal of intellectual enrichment, a society that will achieve a material standard of living as a byproduct of this mentality. Culture and science are seen as goals in and of themselves, not just means to a material end. A high material standard of living is viewed as coming from knowledge and love, not the reverse.

No philosophy of life can logically justify its basic premises. These are givens, the values of the individual or the group. The society that acclaims maximized material consumption as its ultimate goal, that expresses only passing concern for the fate of future generations, that places no value in culture and science other than that which derives from their contribution to consumption, proceeds from a point of reference that cannot be logically overthrown. Such a worldview is the product of an evolutionary process of selection which rewarded clan-specific altruism.

By contrast the eugenics movement advocates a universalism that encompasses all humanity while recognizing the continuity of our species with all other species on this planet, disavowing any exclusively homocentric orientation that would view our fellow creatures as mere fodder for our usage. Eugenicists also perceive a need to be open to genetic manipulation, machine enhancement, and even contact with beings from other planets.

The operative phrase of this ethical system is "the greater good," which is understood more in the spirit of John

Stuart Mill (1806-1873) than in the hedonistic pronouncements of a Jeremy Bentham (1748-1832). The philosophy extends beyond the creature universe to thought itself.

Eugenicists argue that there is much in our genes which may have been advantageous to previous generations and species, but conditions have now changed radically. They maintain that we can either work with nature and achieve utopia, or we can in our greed reject reform and perish. Dangerous? Unquestionably. It is entirely possible, for example, to create people with limited intelligence to perform our manual labor for us, just as we currently import such persons through our national immigration policy. Given our current, still limited understanding, we can easily overestimate our power to predict. And there is the danger of being overly narrow in separating the desirable from the undesirable.

Society and Genes

Politics: Manipulation Masked as Democracy

I believe in the division of labor. You send us to Congress;
we pass laws under which you make money...
and out of your profits, you further contribute to
our campaign funds to send us back again
to pass more laws to enable you to make more money.
Senator Boies Penrose (R-Pa), 1896

There are two things that are more important in politics.
The first is money and I can't remember what the second one is.
Senator Mark Hanna (R-Oh)
Chairman of the Republican National Committee, 1896

In 1999, even as we forged into the new millennium, the Gallup Poll found that 68% of Americans still favored teaching creationism together with evolution in the schools, with 40% favoring exclusively creationism; 47% percent subscribed to the view that "God created human beings pretty much in their present form at one time within the last 10,000 years of so" (up from 44% in 1982!).[57] In the words of the theologian John C. Fletcher, such "controversy clouds rational discussion with fear and misunderstanding."[58]

The genetic bases of social and political structures constitute a topic that even bolder sociologists and political scientists have been leery of raising for two-thirds of a century. It is a taboo which grossly distorts our understanding of ourselves.

There probably has never existed a society with a totally rigid structure in which ability played no role. Under the Caesars, the Pharaohs, the Ottomans, the Tsars, and probably even the Mayan princes, the gifted slave could on occasion demonstrate his ability and achieve high rank. In modern society, however, where such mobility has been immensely increased, universal education combined with assortative mating is creating greater and greater genetic stratification into classes which are then overlaid with stratifications of wealth and power.

In a dictatorship, government is more inclined to determine directly the various functions performed by its citizens, whereas in a democracy the citizenry usually enjoys greater freedom of selection. But even in the most permissive democracy, if the individual does not possess independent means and does not want to starve to death, he must perform *some* function to which society assigns a value. C*ompulsion* is a key word in both systems. This is not stated as a value judgment, but is simply a fact of life. The distinction between democracy and dictatorship has to do primarily with how the authorities get the same tasks accomplished – everything from trash hauling to school teaching – and thus make it possible to maintain a functioning social mechanism and allow those in power to remain in power.

The Skinner box of capitalism has proven to be far more efficient than the Gulag in raising production/consumption. Evidently we have much more in common with cattle than with cats, for we are herded with amazing ease. True democracy is not possible if the people fail to understand the issues. Political history is really nothing more than a broken string of days that will live in infamy.

Dictatorships are difficult to maintain, since a leader who refuses to take account of the disposition of forces in that society will eventually be overthrown. Democracies, on the other hand, possess considerably greater flexibility through manipulation of the popular will.

As for political dialogue, it takes place on three levels: a) sham issues intended to manipulate the masses; b) the true (usually clandestine) views of the ruling elite; and c) long-term species survival issues, which, since the beneficiaries do not constitute a constituency, are generally more ignored than suppressed,.

In 1933, gazing around him in dismay at the Great Depression and peering back at the "holy war fought to make the world safe for democracy," the former civil servant John McConaughy in *Who Rules America?* defined his country's "invisible government" as "the political control for selfish, if not sinister, economic purposes – by individual men, or groups or organizations, who are careful to evade the responsibility which should always accompany power. They operate behind a mask of puppets in politics and business."[59] Exactly a half century later the sociologist G. William Domhoff, whose political views were far to the left of McConaughy's, arrived at similar conclusions in his *Who Rules America Now?* when he described a cohesive ruling class that shapes the social and political climate and plays a dominant role in the economy and the government with the goal of promoting its own self-interest.

No human interaction is more fiercely competitive than politics. What is the true nature of that process? To take but one example, Washington, D.C. is home to a society of "networked," monied, politically sophisticated individuals, while 37% of that same city's residents read at a third-grade level or lower.[60] The situation is comparable to a champion sprinter competing against a 90-year-old in a wheelchair. Not surprisingly, the "winners" in this race favor the process that allows them to achieve and maintain their spoils system, and to do so without any sense of guilt.

One percent of American citizens now own 40% of the nation's wealth.[61] In elections vested interests make electoral campaign contributions, parts of which are used for polling the voters to learn what they want to hear, while the lion's share is invested in advertising that is as based as little on logic as an ad for a soft drink. The resulting advertising presents a combination of what the pollsters discover and what the propaganda specialists consider the populace will accept. To make matters worse, literally a handful of people now control most of the media, and there is no talk of applying antitrust legislation to stop even further amalgamations. And the system functions incredibly smoothly – exactly as intended. When the candidate is eventually elected, having outspent his opponent, he then goes on to do the bidding of those who paid the bill. Should the electoral results be in doubt, the candidate has merely to wrap himself in the flag while denouncing his opponents. The result is an unbridgeable chasm of understanding between elites and the broad masses. A serious book published by a university press may have a print run of a few hundred copies, while a television show of only middling popularity will measure its viewership in the tens of millions, and Hollywood aspires to an audience of billions all over the world. Intellectuals are supposedly free to express their opinions (as least as long as they do not threaten the powers that be), but informed opinion is irrelevant to the political process.

This situation has been made possible by the failure of the general populace to comprehend the true nature of the issues. Indeed, how can any rational observer view any human society as a collective of informed individuals making rational decisions? In a 2000 Gallup poll, 34% of those questioned were unable to name the probable presidential candidates. For persons having a high school education or less and earning less than $20,000 annually, this particular quotient of ignorance rose to 55%.[62] According to a survey done by the National Assessment of Education Progress, 56% of those tested could not correctly subtract 55 and 37 from 100; 18% could not multiply 43 x 67; 24% could not convert .35 to 35%; and 28% were unable to express "three hundred fifty-six thousand and ninety-seven" as "356,097."[63] In addition, 24% of adult Americans were unaware that the United States had fought the Revolutionary War with Great Britain, and 21% had no idea that the Earth revolves around the sun.[64] According to the Northeast Midwest Institute, a nonprofit and education research group, 60 million adult Americans cannot read the front page of a newspaper.[65] Three Americans in ten between the ages of 18 and 24 could not find the Pacific Ocean on a world map, while 67% of Brits did not know the year World War II ended and 64% did now know which country the French Alps were located in.[66]

As for art, philosophy, serious music, literature, and so on – that intellectual thought and creativity which should lend greater meaning to our lives than those of other animals that love, hate, and dream much as we do – such matters are a subject of disinterest for the overwhelming majority of people.

But even this does not represent the furthest extreme of egalitarianist politics. The millions of people ill with dementia to the point that they are unable to dress themselves or recognize family members also participate in selecting national leadership. Surveys of patients at dementia clinics in Rhode Island and Pennsylvania found that 60% and 64% had voted, respectively. Brian R. Ott of Brown University found that 37% of patients with moderate dementia and about 18% with severe dementia had voted.[67]

In selecting out individuals of ability, modern society now has stripped the broad masses of society of the brilliant artisans and poets who formerly created and maintained national cultures.[68] A visit to the magazine section of the local supermarket or a flip through the hundreds of television channels is a dismaying experience.

Welfare and Fertility

See yon blithe child that dances in our sight.
Sara Coleridge, "The Child"

Is the goal of the so-called welfare state fundamentally dysgenic in nature? In 1936, the famous biologist Julian Huxley laid out a hard-hearted version of the hereditarian view in his Galton lecture, delivered before the Eugenics Society:

> *The lowest strata..., allegedly less well endowed genetically..., must not have too easy access to relief or hospital treatment lest the removal of the last check on natural selection should make it too easy for children to be produced or to survive; long unemployment should be a ground for sterilization, or at least relief should be contingent upon no further children being brought into the world.*[69]

We must remember that this was written at the depths of the Great Depression, and that many of those on welfare were simply victims of failed financial policies, not bad genes.

While the average welfare mother receives payments for only two years, never-married mothers who have babies in their teens average eight years or more of dependency.[70] These are the so-called chronic welfare cases. On average the mothers of illegitimate children score ten points lower in IQ than mothers of legitimate children.[71] These babies make an incommensurate contribution to the future pool of rejected, abandoned, and battered children.[72]

The mechanism would appear to be economic. A young woman of average or greater ability can look forward to life's many opportunities and finds little temptation in a modest welfare payment, whereas a woman of low intelligence may rationally see government assistance as a ticket to independence and freedom from the hand-to-mouth realities of a minimum-wage job. It would seem logical that the higher the payments, the greater the temptation. Nonetheless, the link between economics and fertility has been challenged as still unproven. Demographer Daniel Vining, for example, has pointed out that lower welfare payments in southern states has not led to significantly reduced fertility patterns.[73]

We are faced here with a terrible dilemma. Society has an obligation to care for its weakest members, but the flip side of the coin is that in doing so we have significantly increased the fertility of low-IQ women (who generally tend to marry low-IQ men in what is known as "assortative mating"). And we pay them more for each child. Mothers on AFDC had an average of 2.6 children each; non-AFDC mothers averaged 2.1.[74] This is a major factor in American fertility patterns.

What to do? Deny poor women and their children financial assistance? Bribe the upper classes into childbearing? Or throw up our hands in dismay and allow society to be genetically dumbed-down? Indeed, given political realities, what can we do? Certainly, at the very least, it would behoove us to increase family-planning services to the poor.

It is a simple fact that current state policies – both domestic and foreign – already influence differential fertility patterns, despite the fact that the current political climate makes it virtually impossible even to discuss this factor.

Since future generations by definition represent a zero constituency, the public sphere is largely defined horizontally, whereas vertical or longitudinal effects are mostly relegated to the private domain and thus ignored – that is, remain unregulated.

Eugenics opposes this horizontal/vertical opposition, maintaining that, since the unborn constitute a vastly greater potential population than do the currently living, their rights take precedence. Politics is, by definition, a struggle among the currently living, and what may well be a victory for some faction in their midst may well be a disaster for their children, just as the disasters of the parents may be to the children's good fortune.

We are now able to separate sex from procreation; either may occur without the other. It is now even possible for women to bypass the male's sperm.[75] Thus, while leaving the right to sexuality within the private sphere, eugenicists argue that procreational rights – inasmuch as they define the very nature of future people – can be ignored by society only to its own detriment.

Crime and IQ

Oh blood, which art my father's blood,
Circulating thro' these contaminated veins,
If thou, poured forth on the polluted earth,
Could wash away the crime...
Percy Bysshe Shelley, "The Cenci"

Genes play a major role in virtually all behavior, including alchoholism, smoking, autism, phobias, neuroses, insomnia, consumption of coffee (but not tea),[76] schizophrenia, marriage and divorce, job satisfaction, hobbies, and fears. Curiously, while one study shows no genetic role in singing ability,[77] another shows pitch perception to be highly heritable and estimates the heritability of tone deafness at 0.8 – about as high as it gets for genetically complex traits, rivaling features such as height.[78] Animal breeders and even pet owners have no doubts about differences between and within species, and we all know from everyday experience just much people differ innately from each other. Genes evidently also play a role in crime.

In the mid-nineteenth century, criminal justice systems were still guided by the assumption of man's free will, and crime was viewed as a sin which had to be expiated. In the late 1850s, the French physician B. A. Morel established the field of criminal physical anthropology. Galton himself favored compulsory means to limit the breeding not just of the insane, the feebleminded, or confirmed criminals but also of paupers.[79] In 1876, just five years after the appearance of Darwin's *Descent of Man*, the Jewish-Italian criminologist and physician Cesare Lombroso published *The Criminal Man*, which attempted to demonstrate the biological nature of criminality. Lombroso claimed to have established during autopsies certain physical stigmata characteristic of the born criminal, whom he saw as possessing a more primitive type of brain structure. If one accepts such biological determinism, punishment becomes meaningless.

Lombroso's theories are now generally rejected as invalid, but studies of the role of genes in crime have not been confined to the nineteenth century. A 1982 Swedish study found that the rate of criminality in adopted children was 2.9% when neither biological nor adoptive parents had been convicted of criminal activity. When one of the natural parents was criminal, the figure rose to 6.7%, but when both biological parents were criminal, the figure was nearly twice as high – 12.1%.[80]

At first the left tended to sympathize with biological positivism, but soon Marxists came to view crime as environmentally determined. The anarchists even sympathized with criminals, who were seen as rebels challenging social injustice. Crime in a capitalist system came under the rubric of justified revolution in miniature.

If the egalitarian Franz Boaz was the "father" of anthropology, the paternal rights to criminology (sociology's "stepchild") have been ceded to Edwin E. Sutherland, for whom learning was entirely a social product disconnected from biological structures. In 1914, he published *Criminology*, the most influential book on the topic during the twentieth century. Thanks in large measure to its resonance, and especially that of later reworked editions, many textbooks in the field never even mentioned IQ, and when they did the treatment was largely dismissive.

At the same time, intelligence studies have consistently demonstrated a lower IQ among those found to have committed criminal acts than among the general population. The intelligence ratings of 200 juvenile offenders consigned to training schools in Iowa show a mean IQ of 90.4 for the boys and 94.1 for the girls. The mean IQ for nondelinquents was 103 for boys and 105.5 for girls.[81] The 1969 police records of over 3,600 boys in Contra Costa County, California, show a relationship between IQ and delinquency of -0.31.[82] A group of 411 London boys was followed over a ten-year period so as to compare delinquent and non-delinquent groups. While only one in fifty boys with an IQ of 110 or more was a recidivist, one in five of those with an IQ of 90 or less fell into this category.[83] Since the advent of the revised Stanford Binet and the Wechsler-Bellevue scales in the late 1930s, it has been consistently found that samples of delinquents differ from the general population by about 8 IQ points[84] – a significant but not an overwhelming difference. One can only surmise that perhaps the gap would be even narrower if it were possible to control for a higher arrest record among juveniles less skillful in the art of deception. The same general tendency exists within the adult population. Criminal offenders have average IQs of about 92 – that is, 8 points or one-half standard deviation below the mean.[85]

What is actually happening? Life itself is a cruel competition, where the vanquished have ended up more than once skewered and slowly roasting over the victor's cooking fire. Now civilization imposes rules (so-called middle-class values) that allow some people more success at winning. Imagine a situation where the fastest runner would be the only one to get supper. After a time the slower competitors would be sorely tempted simply to hit him on the head rather than futilely attempt to outdo him in speed. The same is true with intelligence. The successful stockbroker, surgeon, and lawyer do not need to commit crime to gain wealth, but further down the professional scale are those individuals whose low intelligence literally dooms them to a life of material slavery. Can at least part of the explanation for criminal behavior be as simple as that?

To what extent is inherited low altruism a factor in crime? Before axing the old pawnbroker in Dostoevsky's *Crime and Punishment*, Raskolnikov first rationalizes away his guilt. Clearly, the general population contains a vast pool of individuals for whom guilt is, at best, an underdeveloped emotion.

Can we really entrust the awesome task of guiding human evolution to the bureaucrats? Are we not still far from understanding the nature of crime? Do we want passivity bred into the population? Is not crime the statistical tail of such desirable traits as adventuresomeness and the willingness to take risks?

Migration

Settling and dominating the entire planet, our species has devoted an immense amount of effort to moving around. In the process, entire civilizations have been displaced, conquered, infiltrated, and even swamped by imported alien populations. In economic terms, greater and greater specialization has replaced self-sufficiency and created ruling classes that are often recruited from a multiplicity of ethnic backgrounds.[86]

Since the pool of global talent is neither diminished nor enhanced when a person moves from country A to country B, migration constitutes a zero-sum game. Nevertheless, some countries are winners while others are losers. The United States attracts large numbers of very talented individuals but also many who are unlikely to leave the lower economic rung. The mean IQ of immigrants in the 1980s has been estimated to be about 95, or only about one-third standard deviation below the mean.[87] This is a small enough difference that it may well be explainable by the disadvantaging environment from which many arrivals come.

Early man migrated slowly, creating diversity by virtue of lengthy periods of relative genetic isolation. Now, however, the revolution in transportation is undermining this isolation. The United Nations Educational and Cultural Organization (UNESCO) estimates that 53% of the 6,809 languages spoken around the world are at risk of extinction by 2100. The destruction of this "reservoir of human thought and knowledge"[88] is accompanied by a loss of genetic diversity that would cause dismay among ecologists if it were to occur in any species other than man.

The History and Politics of Eugenics

A Brief History of the Eugenics Movement

The first stages of plant and animal-breeding mark the end of the hunter-gatherer period of human evolution. As far as written testimony is concerned, Plato's *Republic* provides an early theoretical treatise on eugenics.

Once Darwin's 1859 *Origin of Species* had established both the mechanism of evolution and man's place in nature's greater scheme of things, it was inevitable that people would want to engage in what was then referred to as "racial" improvement. They would, at the same time, worry about the genetic consequences of eliminating natural selection in the modern world. Darwin himself became a true Social Darwinist, bemoaning the fact that:

> *We do our utmost to check the process of elimination; we build asylums for the imbecile, the maimed, and the sick; we institute poor-laws; and our medical men exert their utmost skill to save the life of every one to the last moment.... Thus the weak members of civilized societies propagate their kind. No one who has attended to the breeding of domestic animals will doubt that this must be highly injurious to the race of man.*[89]

It was Darwin's cousin, Sir Francis Galton, who in his 1883 book *Inquiries into Human Faculty* coined the word "eugenics." Even earlier he had done pioneering work in his *Hereditary Genius* (1869) and *English Men of Science: Their Nature and Nurture* (1874). Galton was also one of the first to recognize the importance of twin studies. He also proved to be correct (unlike his more famous cousin) in rejecting the Lamarckianism of the age, which held that acquired characteristics could be passed on to offspring.

In 1907, the Eugenics Education Society was founded in London, and eugenics enjoyed broad support among the British elite, including that of Havelock Ellis, C. P. Snow, H.G. Wells, and George Bernard Shaw. The last wrote that "there is now no reasonable excuse for refusing to face the fact that nothing but a eugenics religion can save our civilization from the fate that has overtaken all previous civilizations."[90]

The movement was also strong in the United States. In the 1870s, Richard Dugdale published his famous study of the Juke family, unearthing 709 members of a single family with criminal pasts. By the 1880s, custodial care was widely introduced to prevent the feebleminded from reproducing, and by the end of the century, there were cases of sterilization of the feebleminded. 1910 saw the founding of the Eugenics Record Office at Cold Spring Harbor, on Long Island. Alexander Graham Bell, who was wed to a deaf woman and was concerned about the interbreeding of the deaf, feared that such selective mating could lead to the creation of a deaf population. He became a prominent member of the American eugenics movement.

The influence of the eugenics movement did not derive from the number of its members. Both in Great Britain and in the United States adherents numbered only a few thousand. Rather, the influence of the movement was explained by the wealth and influence of an elite and, unfortunately, an often elitist group.

After 1910, eugenics societies were founded in various American cities, and a number of Americans attended the First International Eugenics Congress in London in 1912. The Second and Third were held in New York, in 1921 and 1932, respectively.

When World War I broke out, eugenicists helped the U.S. Army develop intelligence testing, and they proselytized widely after the war. In the 1920s, they played a major role in tripling the number of institutionalized feebleminded and in vastly increasing extra-institutional care.[91] As for sterilization, contrary to popular belief, eugenicists were split down the middle on the issue. Neither the National Committee for Mental Hygiene nor the Committee on Provision for the Feebleminded supported sterilization.[92] Part of the reason for the reluctance was that eugenicists were a straight-laced lot, who were afraid that sterilization could lead to a loosening of sexual mores. Neither, for that matter, were they particularly eager to see eugenics tarred with the polygamist brush.

By 1931, 30 states had passed a sterilization law at one time or another. Even so, the number of actual sterilizations was modest on a national scale. By 1958, these amounted to only 60,926.[93] In comparison, twenty million sterilizations were performed in India between 1958 and 1980, and in China some thirty million women and ten million men were sterilized between 1979 and 1984. An undetermined number of these were coerced.[94]

German submarine warfare had temporarily braked free immigration to the United States during World War I. In 1924, Congress was strongly influenced by eugenic considerations in framing immigration law, so that immigration flows were made to reflect the ethnic makeup of the country as a whole. On July 1, 1929, national origin quotas were established as the basis of American immigration policy.

The subsequent history of eugenics is presented in the next four subchapters. We can note here only the enormous current interest in the topic. A search of the Online Computer Library Center (OCLC, or "Worldcat") on the World Wide Web revealed some 3,200 published books on the topic. Eighty-four of them preceded Galton's 1883 coinage of the word:

OCLC Search for Books on Eugenics

before 1883	84		1940-1949	243
1883-1889	14		1950-1959	128
1890-1899	23		1960-1969	138
1900-1909	124		1970-1979	146
1910-1919	536		1980-1989	230
1920-1929	419		1990-1999	396
1930-1939	569		2000-2005	452

If visual and sound recordings are added to the 2000-2005 book search, the number comes to 610 – greater than the annual average for books during the peak period of 1910-1919. Given the revolutionary progress of the science of genetics, it is a safe bet that this trend represents a rising curve. There is also a flood of articles on eugenics circulating over the Internet – a medium nonexistent in 1910-1919. A January 2006 Internet search for eugenics using Google produced 1,840,000 items as opposed to 231,000 as of April 2004. Thus, the popular view of eugenics as a bygone historical phenomenon is patently incorrect.

Germany

Eugenics is now popularly presented as the ideology of Holocaust and, as such, is an object of intense vilification. Leo Strauss, the philosopher and Zionist member of the Jewish Academy, coined the maxim "reductio ad Hitlerum": *Hitler believed in eugenics. X believes in eugenics. Therefore X is a Nazi.*[95]

It is impossible to discuss the eugenic platform without treating the history of eugenics in Germany. To do so we must begin farther back in time than the period of 1933 to 1945.

During the late nineteenth century the upper classes in Germany – and not only in Germany – turned to Social Darwinism as a justification for the disproportionate wealth which they had accumulated. Thus it was no surprise that in 1893 Alexander Tille promoted the idea that a people which has been raised in the consciousness of competition as a mechanism for achieving progress "will be difficult to convert to Socialist daydreams."[96]

Aside from economic class, race was a much abused theme. The subject of degeneration in animals had been raised by the French naturalist Georges Buffon (1707-1778) in 1766, and as early as the 1820s the topic had drawn broad public attention. The French Count Joseph de Gobineau (1816-1882) developed the notion still further, applying it to humans and postulating the existence of an "Aryan" race that supposedly formed the basis of "Nordic" populations. The last remaining Aryan groups were seen by him as inhabiting Northern Germany and England. According to Gobineau, the interbreeding of Nordic types with other groups would lead to degeneration. Gobineau was best received in Germany.

In 1895, the German amateur anthropologist Otto Ammon preached a gospel of interbreeding "the pure original type with somewhat dark long-skulled types and round-skulled types with somewhat lighter pigment. All intermediate mixed forms do not count among the great successes, but are given over to the struggle for existence, for they were created only as inevitable byproducts in producing the better."[97]

A relatively small group of German physicians, some of whom were related to each other by marriage, picked up on Galton's eugenics and degeneration – but from a leftist point of view. The founder of German eugenics, Alfred Ploetz (1860-1940), was a socialist. In 1891, Wilhelm Schallmayer (1857-1919) published a brochure on species decline, but, while Galton's interests related largely to intellectual abilities, Schallmayer was captivated by the idea of physical degeneration. Schallmayer maintained that Darwin, having discovered the causal nature of evolution, thus rendered that process manageable. Schallmayer was opposed to Gobineau's racial theories. Alfred Grotjahn (1869-1931) concurred that there was a danger of genetic decline and saw the theory of degeneracy as an important step in the process of "medicalizing" the problem.

The theses of the German Society for Racial Hygiene, adopted in 1914, stood in marked contrast to Gobineau's views and made no mention of either class or race. (The phrase "racial hygiene" was coined by Ploetz in 1895 as an alternate name for eugenics. Its use was unfortunate in that it often came to be misinterpreted as referring to individual races rather than to the human race as a whole.) The theses called for family-friendly housing; elimination of factors that might hinder members of certain male professions from having children; raising the taxes on alcohol and tobacco; legal regulation of medically required abortions; combating what was then viewed as the hereditary transmission of gonorrhea, syphilis, tuberculosis, and diseases acquired in the course of practicing a profession; mandatory exchange of health certificates prior to marriage; and the awarding of prizes for literary and art works in which family life was praised. Young people were asked to be ready to sacrifice for the communal good.[98]

By the end of the 1920s eugenics had moved beyond the small group of specialists to become a topic of national discussion. The Society's 1931/32 theses again stressed the importance of inheritance, warned of degeneration, and stressed the importance of the family, calling for a heightened birthrate and the provision of tax relief for families. Lengthy periods of professional training were recognized as undermining fertility, genetic counseling was recommended, childbearing by persons whose children were likely to suffer from genetic illness was to be discouraged, and young people were to be instructed as to their eugenic obligations to their children.[99] Once again, no mention was made of race.

Nineteenth-century Social Darwinists had viewed war as an invigorating process that weeded out the weak, just as economic competition sorted out a population into classes according to fitness. As World War I dragged on, eugenicists came to judge it "counter-selectionary."

Prior to the end of World War I there had been a real fear in Germany of overpopulation. The population of the German empire had grown from 45 million in 1880 to 67 million by the end of the First World War. Only in 1918-1919 did the number of deaths exceed the number of births.[100] The new fear of underpopulation made it more difficult to propagandize negative eugenics, but "racial hygienists" attacked the Malthusians on the grounds that precisely the more desirable elements of the population were most likely to heed their calls for restraint and that this ill-advised altruism would prove to be dysgenic. They were also concerned that population decline would pose an existential threat to the "Nordic race." Within the context of theories of racial superiority, racial interbreeding was seen as a sort of suicide of those of the "superior" race.

Nevertheless this was not what originally concerned Adolf Hitler. In 1920, he put forward a list of 25 points, none of which dealt with eugenics. The word "eugenics" never even appears in *Mein Kampf.*

To best comprehend the role of eugenics under the National Socialist government, and not limit my examination of German eugenics to a narrow context, I approached the topic by first selecting one hundred books dealing with the Weimar and Nazi periods which contain indexes covering not only proper names but topics as well. I made no attempt to preselect other than choosing volumes that deal with the period. All hundred books are listed in Appendix 2. It is an experiment that anyone with an afternoon to spare and access to a serious library can easily replicate, selecting whichever books he or she might like.

The authors of these books range from Nazi ideologues to recognized Western scholars. Ninety-six of these indexes did not contain the word "eugenics." The four volumes whose indexes listed eugenics contained only a handful of mentions. Even the indexes to *Mein Kampf* and Hitler's speeches do not list eugenics as a topic, although they contain numerous references to race. Obviously, eugenics was not the powerful ideological motor it is made out to be.

Still, Hitler had heard of eugenics and eventually came to view it – approvingly – as being of a single piece with his ideas of Social Darwinism and a mystical "Nordic" or "Aryan" race, much in the spirit of Gobineau (whose name is never mentioned in *Mein Kampf*). This was a case of explicit tribalism buttressed with superstitions and mysticism, eventually even producing expeditions to the Himalayas in search of roots, and the prominent use of Germanic pagan symbols and runes.

While Hitler may have been a dyed-in-the-wool hereditarian, he was also an anti-universalist who saw the production of a pure Nordic stock as the ultimate goal of genetic selection. Rather than view the development of humanity as one of cooperation, he held to a doctrine of competition. Abilities displayed by other peoples were for him negative phenomena which threatened the group he proposed to champion. This anti-universalist system of values represented a system of values that was anti-eugenic in the most fundamental sense.

A number of German eugenicists held views opposed to the government's vision of "racial hygiene." Hans Nachtsheim, a proponent of voluntary sterilization and Germany's leading geneticist after the conclusion of World War II, consistently rejected the Nazis' ideas of race. Even Fritz Lenz, who was perhaps the most influential German eugenicist during the Nazi period, spoke out against anti-Semitism. The biologist and eugenicist, Professor Walter Scheidt, denounced the unscientific nature of "racial biology" as taught at German universities. Still another proponent of eugenics, the Viennese physician Julius Bauer rejected Nazi concepts of race as "fantasies plucked from the air" and complained bitterly as to the harm they were doing the cause. A fellow Austrian physician and supporter of eugenics, Felix Tietze, condemned the Nazi law on "Protection of the Blood." The biologist and eugenicist Juliux Schaxel protested the exploitation of eugenics by the Nazis and actually emigrated to the Soviet Union. Rainer Fetscher and the former Catholic priest Hermann Muckerman were dismissed from their positions because their worldview contradicted that of the Nazis, and Fetscher ended up being shot by the SS when he attempted to make contact with the Red Army.[101]

Eugenicists in other countries explicitly rejected Hitler's anti-Semitism and racism. At the International Eugenics Conference held in Edinburgh in 1939 British and American geneticists criticized the racist orientation of eugenics in Germany.[102] That same year prominent eugenicists in the United States and England issued a statement explicitly rejecting "race prejudices and the unscientific doctrine that good or bad genes are the monopoly of particular peoples" (see Appendix 1).

But the National Socialist government took control of scientific institutions and funded a number of chairs of "Racial Hygiene" in German universities, so that eugenicists abruptly found themselves face to face with the temptation to leave behind the pack of daydreaming social reformers and begin to implement eugenic reform.

One geneticist who became an ideologue of Nazi crimes was Otto von Verschuer. His essay, "The Racial Biology of Jews," appeared in Hamburg in 1938 as one of nearly fifty articles, published in six volumes, under the title *Forschungen zur Judenfrage* (Studies on the Jewish Question). The research had been subsidized by the National Socialist government.

The article purports to treat physical differences between Central-European Jews and Germans. Verschuer points out the astonishing phenomenon that an ethnic group could preserve itself for two thousand years without a territory. He then goes on, quite correctly, to point out that the differences he describes are not absolutely applicable to either group but are a matter of relative frequency within the two groups. Taking a great deal of trouble to impart a scientific tone to the text, including such characteristics as, for example, fingerprints, blood types, or vulnerability to specific diseases – all of which pose fully legitimate questions for the physical anthropologist – he nevertheless presents a pathological document of ethnic hatred disguised as science. The Jews, we learn from Verschuer, have hooked noses, fleshy lips, ruddy light-yellow, dull-colored skin, and kinky hair. They have a slinking gait and a "racial scent." Verschuer then moves on to "pathological racial traits." He does concede high intellect and a relatively low birth rate, but by the end of the article his hatred becomes blatant:

> *I believe that only people of a certain type feel attracted by Judaism and could decide on conversion to it, people in particular who felt related to Judaism on the basis of their intellectual and psychological makeup. (It may only seldom have been physical reasons.) In this sense, the element which was absorbed in Jewry was not "foreign."*

Verschuer then goes on to conclude that there is an absolute necessity for Germans and Jews to remain separated. It was a position identical to that laid out in *Mein Kampf*, whose author states that "the most lofty human right and obligation is to preserve the purity of the blood." Once that primary task has been accomplished Verschuer then insists on combating childbearing by "syphilitics, persons suffering from tuberculosis, persons suffering from genetic disabilities, cripples, and cretins."[103] That is, he is first and foremost concerned with the prevention of interbreeding with other groups, and only after that with disability, heritable or nonheritable.

Although nowhere in the article does Verschuer use the word "eugenics," he saw his argument as being fundamentally "eugenic." It is, after all, so convenient for someone consumed with hatred to claim his arguments are the product of scientific reasoning and not emotion. True, he does not call for an extermination of the Jews, but the train of his logic is very close to doing precisely that. Verschuer was a mentor for Joseph Mengele, who was keenly interested in twins research.

There is probably nothing in the universe that cannot be twisted, distorted, and used for evil. The danger of the misuse of science will always be with us. It is even more disheartening to see that this product of either a sick mind or shameless opportunism has been translated and distributed by a translator who displays a Ph.D. after his name.

Verschuer's *Manual on Eugenics and Human Heredity* was published in French translation in German-occupied Paris in 1943. His signature on the preface is dated summer 1941. Much of the book contains the facts of heredity, as known at the time, a statistical distribution of variance, and so on, and is simply a popularized textbook on human genetics. In it he writes that the prominent eugenicists Erwin Baur, Eugen Fischer, and Fritz Lenz all read the manuscript and made suggestions.[104] Obviously, to make the document acceptable to them, he avoided the insidious anti-Semitism of the earlier essay, maintaining that "Galton's eugenics and Ploetz's racial hygiene were in complete agreement with regard to both content and goal."[105] He also praised Gobineau's *Essai sur l'inégalité des races humaines*. Darwin, Mendel, and Karl Pearson were also praised as pioneers of eugenic thinking.

*

There are three basic charges associated with eugenics under National Socialism: a) the July 1933 sterilization law; b) the September 1939 national euthanasia program; and c) the persecution of Jews and gypsies and their mass murder toward the end of the war. Let us examine each in order:

A bill was drafted in 1932 by the Prussian Governmental Council – *before Hitler's accession to power* – to lay the groundwork for selective sterilization in cases of heritable diseases. Although sterilization had been discussed for twenty years, the legislation took the leading German eugenicists by surprise, who were critical of it as counterproductive and inefficient with regard to genetic improvement.[106] On July 14, 1933, the legislation was passed by the German parliament, entering into force in 1934, but now it permitted sterilization against the wishes of the individual concerned, specifically for the surgical sterilization of persons whose offspring would have a high probability of suffering from physical or mental illness, of hereditary feeble-mindedness, schizophrenia, manic-depressive syndrome, hereditary epilepsy, Huntington's disease, hereditary blindness, deafness, or severe physical defects, as well as severe alcoholism.[107] No mention was made of race. From 1934 to 1939 an estimated 300,000 to 350,000 persons were sterilized.[108] Most sterilizations were for feeble-mindedness, followed by schizophrenia.[109] At the time, sterilizations were also being practiced in a number of European countries and the United States, although on a smaller scale. Eugenic considerations did not play a significant role in the debate. Rather, German legislators misguidedly saw sterilization as a cheap alternative to welfare.[110] The Catholic Church was opposed to sterilization, but the Evangelical Church supported it.[111]

The debate over euthanasia was launched by Karl Binding and Alfred Hoche's 1920 book *Legalizing the Destruction of Life Not Worth Living*. The authors, a lawyer and a physician, put forward a strictly economic argument. While there may have been some peripheral eugenic case to be made for the sterilization legislation, the euthanasia question had nothing whatever to do with eugenics, since persons who were already institutionally segregated and in many cases sterilized could not have had any procreation. To their credit, German eugenicists vehemently attacked euthanasia proposals. In 1926, the eugenicist Karl H. Bauer, for example, stated that if selection were used as a principle for killing people, "then we all have to die"; the eugenicist Hans Luxenburger, in 1931, called for "the unconditional respect of the life of a human individual"; in 1933, the eugenicist Lothar Loeffler argued not only against euthanasia, but also against eugenically indicated pregnancy terminations: "we justifiably reject euthanasia and the destruction of *life not worth living*."[112] Hitler, however, regarded the institutionalized as "useless eaters" who were taking up the time of hospital personnel and occupying bed space to no worthwhile purpose.[113] When, in September 1939, he issued a secret order initiating a national euthanasia program, he did so strictly to free up as many as 800,000 hospital beds for expected war casualties.[114]

The murder of huge numbers of Jews is an undeniable fact, but it is not accurate to regard the eugenics movement as the ideological engine of this Holocaust. It is true that Hitler, partly under the influence of a manual on human heredity and eugenics written by Erwin Baur, Eugen Fischer, and Fritz Lenz, supported eugenics,[115] but he did not hate the Jews because he had been taught by eugenicists to classify them as intellectually inferior. On the contrary, he regarded them as powerful competitors of the blue-eyed, blond race he proposed to champion. The Jews were blamed for Germany's defeat in World War I and for the humiliations of the Versailles treaty. When it became apparent that a new defeat awaited Germany as a consequence of World War II, vengeance became the order of the day. As for the gypsies and Slavs, the former were to be exterminated and the latter could be exploited as slaves captured from an inferior tribe. The mass murders of Jews, gypsies, and many Slavs during the late war period took place in absolute secrecy. The community of German eugenicists did not call for a holocaust.

Nevertheless, it is equally undeniable that there were German eugenicists who allowed themselves to be co-opted by the regime and who helped to create a climate of legitimization of policies of hatred for other ethnic groups. By giving themselves over to ethnic partisanship rather than universalism, they harmed not only the specific victims of Nazi atrocities but their own system of values and beliefs.

Intellectual history is replete with instances of idealism taking disastrous turns. Christianity and socialism must forever bear their respective crosses of Inquisition and Gulag. Eugenics is not the ideology of Holocaust, but in one specific country a small group of its adherents, a group that had already shrunk even further in the changing climate of contemporary genetics, was guilty of complicity. Nevertheless, this was not the driving force behind National Socialism that it is popularly made out to be. Rather, eugenics was an argument that could be conveniently twisted by the Nazi government over the explicit objections of the movement's leaders.

Left and Right

Remember,
every step to the right
begins with the left foot.
Aleksandr Galich (Ginzburg)

While there was a definite association between Social Darwinism and *laissez-faire* capitalism, the debate on eugenics actually cut across class and political lines throughout Europe and America, and it is historically incorrect to associate the movement exclusively with the political right. To no small degree it grew to prominence as part of a search for an exit from the excesses of unbridled nineteenth-century capitalism. Even when Herbert Spencer, in England, and William Graham Sumner, in the United States, began defending the period's gross social inequalities, the left was not about to renounce natural selection, and proponents of socialism saw no inherent contradiction between the two schools of thought. Marx and Engels were themselves enthusiastic Darwinists, feeling that the theories of evolution and communism were mutually complementary sciences that dealt with related but different topics – biology and social interaction. Vladimir Lenin himself derided the claim that people are equal in ability.[116] Galton's chief pupil and the leader of Britain's eugenics movement, Karl Pearson, was a Fabian socialist, as was Sidney Webb, who contributed an essay on eugenics to the influential 1890 *Fabian Essays*. Geneticists in the early Soviet state attempted, unsuccessfully, to model the socialist experiment along eugenic lines.

There was an influential "Weimar Eugenics" prior to Hitler's ascent to power in Germany, where eugenics and socialism were viewed as mutually complimentary – a symbiosis that is still difficult for today's left to accept.[117] The "father" of German eugenics, Karl Ploetz, was a socialist who even spent four years in the United States exploring the possibility of establishing a socialist pan-Germanic colony there. The Austrian feminist and socialist journalist Oda Olberg, who went into exile during the Nazi period, was keenly interested in the ideas of Wilhelm Schallmayer, who attempted to achieve a fusion of eugenics and socialism and vigorously opposed all forms of racism. Another of Schallmayer's fans was Eduard David, one of the leaders of Social Democrat Revisionism. Max Levien, head of the Munich chapter of the German Communist Party, wrote that eugenics would play a role in the development of humanity as a function of technical progress.[118] Alfred Grotjahn favored efforts, within a socialist framework, to reduce the birthrate of the genetically disadvantaged, and the influential socialist theoretician Karl Kautsky took degeneration for granted. There was even a considerable eugenics faction in the Social Democrat Party.

In the heyday of eugenics, the geneticist H. J. Muller argued that the privileges of capitalist society too often promoted persons of limited ability and that society "needed to produce more Lenins and Newtons."[119] Another confirmed Marxist, the distinguished geneticist J. B. S. Haldane, commented in 1949 in the *Daily Worker* that "The formula of Communism: 'from each according to his ability, to each according to his needs' would be nonsense, if abilities were equal."[120] The geneticist Eden Paul summed up the view of many on the left: "Unless the socialist is a eugenicist as well, the socialist state will speedily perish from racial degradation."[121]

The traditional breakdown between *left* and *right* can be fundamentally rephrased as "redistributive" and "competitive," respectively. Logically, egalitarianism is consistent with the *competitive* point of view. If we are really all "equal," we should for consistency's sake favor a "best man wins" approach. If, on the other hand, inequality is genetically preprogrammed, then fairness demands that *redistribution* become the order of the day, first of material goods, and – with time – of genes. Eugenicists point out that if a material good can, by definition, be redistributed only by confiscating from one person to give to another, genetic redistribution does not suffer from this zero-sum limitation.

Holocausts were supposed to have been the creations of hereditarians, not egalitarians, but the left has generally discredited itself no less than the right with its mass murders. And then, too, there was the ubiquitous economic collapse of socialist economies, the self-serving tyranny of their bureaucracies, and the poverty into which they had managed to drive their own populations. It is not a good time for leftist ideology, and self-examination is definitely on the agenda – on the most fundamental level.

As the second millennium came to a close, Yale University Press published a tiny volume by the bioethicist Peter Singer, who attempted to bridge the gap between leftist political thought and Darwinism. Singer propounds a socialism based on championing the rights of the downtrodden. He points out that the 400 richest people in the world possess a combined net worth greater than the bottom 45%. He takes up their cause, arguing that it was the political right that had attempted to co-opt Darwinism, while the left made the mistake of accepting the right's assumptions. "It seems implausible," Singer maintains, "that Darwinism gives us the laws of evolution for natural history but stops at the dawn of human history.[122]

In principle, Singer is correct in maintaining that a "Darwinian left" can again arise, although traditional Marxists who regard their founding father as a prophet-like figure whose views have forever determined what is left and what is right will undoubtedly point out his famous dictum that "social being determines consciousness." And Marx was, it should be mentioned, hostile to Malthusian thinking, which has often gone hand in hand with eugenics and the right-to-die movement.

The notorious nature/nurture debate has been grossly exaggerated by sophisticates who in reality are far less "egalitarian" and "environmentalist" than they would have their naïve followers believe. The true conflict rages between interventionism and a *laissez-faire* approach. If one imagines a continuum with hereditary factors at one end and upbringing at the other, there are three basic possible positions which one can take:

- genetic determinism explains the diversity between individuals and groups, with environmental factors playing a trivial role;
- environmental conditioning overwhelms any genetic predispositions;
- hereditary factors and environmental conditioning interact.

In reality, unalloyed genetic determinism is partly a memory of nineteenth-century social Darwinism and partly an invention of egalitarian environmentalists, who attribute such views to their opponents in an attempt to discredit them. As for the all-nurture school, it remains a lovely fantasy (would it were true!), which all but the most radical egalitarians have abandoned. There is only one tenable view of nature/nurture – that of interaction, not mutual exclusion. Legitimate differences of opinion relate only to the relative importance of the one factor *vis à vis* the other.

Egalitarians have erected a multiplicity of arguments:

a. Modern man represents a *tabula rasa*, a clean slate upon which environment can write any text.
b. There are no significant intergroup differences.

c. While differing levels of individual skills may exist on an intragroup basis, there is no such thing as general intelligence.
d. IQ tests do not test intelligence but only the ability to take tests.
e. The heritability of intelligence is zero.
f. Even if one concedes that the fertility patterns of modern society are dysgenic, evolution does not always follow Darwin's gradualist model, in which minor alterations lead over time to major evolutionary changes. Rather a "punctuated equilibrium" governs lengthy periods of genetic stasis. This seemingly scientific argument, applied, for example, to crustaceans, is a true Trojan horse really intended to be dragged into the gates of the human city.

The foregoing are essentially delaying tactics, but they have created in the public mind an assumption of genetic exclusionism – the assumption that humankind has emancipated itself from subsequent evolution.

Ultimately science cannot be stopped by historical events, however tragic they may be. University of Massachusetts political scientist Diane Paul has summed up the current intellectual climate quite well:

> *Virtually all of the Left geneticists whose views were formed in the first three decades of the century died believing in a link between biological and social progress. Their students, coming to intellectual maturity in a radically different social climate, either did not agree or, in a social climate inhospitable to determinism, were unwilling to defend that position. The appearance of sociobiology probably signifies a fading of the bitter memories surrounding the events of the 1940s. As those memories recede, it would not be surprising to witness the reemergence of a doctrine that was never defeated in the scientific arena but rather submerged by political and social events. From the late 1940s to the early 1970s, it has been, perhaps, a viewpoint latent among scientists only requir-*

> *ing another change in the social climate to prompt its expression.*[123]

Biologist Lawrence Wright, basing his assessment on the University of Minnesota twin studies, concludes that

> *The prevailing view of human nature at the end of the century resembles in many ways the view we had at the beginning.*[124]

Because of the heated nature of the debate, the ideological lines of the various participants often appear fuzzy to the observer, and, on occasion, even to the participants. Below are laid out four basic positions, two of which are egalitarian – "naïve egalitarianism" and "sophisticated anti-interventionism." The reason for the latter distinction is that sophisticated egalitarians are in some respects in greater agreement with eugenicists than with naïve egalitarians. Naïve egalitarians may claim to be adamantly opposed to eugenics but are able to define the concept only vaguely or perhaps not at all. Basically, sophisticated egalitarians are leery of revealing or discussing their own true views for fear of a possible misuse of genetic knowledge.

The following chart has a certain artificiality to it, since people do not fit into neat, distinct groups. National Socialism, for example, attempted to erect a eugenic superstructure over a Social Darwinist base.

	Eugenics	Social Darwinism	Naïve Egalitari- anism	Sophisticated Anti- Interventionism
Universalist/Tribalist	Universalist	Tribalist	Universalist	Mixed
Human evolution	Admit	Admit	Mixed ad- mis- sion/denial	Admit
Natural selection of humans	Oppose	Favor	Oppose	Oppose
Artificial selection of humans	Favor	Mixed favor/oppose	Oppose	Oppose
Current intragroup diversity	Admit	Admit	Either deny or admit but deni- grate	Privately admit but publicly deni- grate
Current intergroup diversity	Admit	Admit	Deny	Privately admit but publicly deny
Intragroup selection	Feasible and desir- able	Feasible and desir- able	Neither feasible nor desirable	Feasible but too dangerous
Intergroup selection	Feasible but not desirable	Feasible and desir- able	Neither feasible nor desirable	Feasible but not desirable
Future intragroup diversity	Admit	Admit	Mixed ad- mis- sion/denial	Privately admit but publicly deni- grate
Future intergroup diversity	Feasible and desir- able	Feasible but not desir- able	Deny (not feasi- ble)	Feasible and Desirable, but not essential
Long-term group coexistence	Desirable	Not desir- able	Desirable	Desirable

Aside from conflicting ideologies, a huge range of sophistication also exists within the various camps. The following is a simplified breakdown by group:

Social Darwinists. Although they were major players in the second half of the nineteenth century and the first half of the twentieth, they have lost their viability as a distinct group. Selection by mortality has been overwhelmed by selection through fertility, although epidemics such as AIDS and modern warfare may one day reverse this equation, possibly sooner than we think. Nevertheless, Social Darwinism still exists as a "residual" philosophy embedded in the very core of the ideologies of certain groups.

The "Nordic" or "Aryan" idea. Driven underground as much by the Holocaust memorial movement (in which the author of this book played a modest role), which was launched after the 1967 Arab-Israeli war, this group has been reduced to arguing for white survival rather than for white supremacy. The average woman in Europe now bears only 1.4 children, whereas 2.1 are needed just to maintain a population. According to the Population Reference Bureau's *2005 Population Data Sheet*, the population of Europe will drop from 9.8% of the global population to 6% by 2050, despite projected strong in-migration. Equally ominous to these theoreticians are the genetic consequences of racial interbreeding inevitable in the "global village." This group's loyalties are drawn along ethnic lines, not class. They can be termed tribalists.

Sophisticated anti-interventionists. This is a group which opposes intervention in the human germ line, and some of its members are opposed to intervention even in the germ lines of animals and plants. The anti-interventionists were traumatized by the German slaughter of Jews and by the lip service paid by the National Socialists to eugenics, and this circumstance has shaped their views accordingly. Strangely enough, the private position of this group has much in common with that of the eugenicists. There is a considerable gap between the group's core beliefs and the views which it proselytizes. It wields influence vastly incommensurate with its size. Some sophisticated anti-interventionists are actually tribalists.

Naïve environmental egalitarians are people who have not given much thought to population and who have accepted the mass-consumption egalitarian gospel disseminated by the anti-interventionists. The goal of any propaganda campaign is to achieve a "disconnect" from practical experience in the targeted population, and in the case of naïve egalitarians this goal has been admirably achieved. They accept that intelligence is strictly the result of education and that altruistic behavior or the lack of it is exclusively the result of upbringing. They reject even the theory of evolution.

Universalist eugenics is described in this book in some detail, so that a description at this point would be repetitive. Suffice it to say that eugenicists see themselves as a lobby for future generations.

Neo-Malthusians. As many nations pass through the demographic transition, this group is losing much of the credence it enjoyed only recently. Most demographic forecasts now predict a leveling off of global population growth, but the Malthusians argue that the population may well be too large already to be self-sustaining and that rapid population growth is still alarming in many areas of the planet. Most eugenicists tend to be Malthusians, but the reverse is not necessarily true.

Anti-Malthusians. This group maintains that human capital is itself the greatest resource and that fears of exceeding the planet's "carrying capacity" are grossly exaggerated and misplaced. In theory, eugenicists could conceivably be anti-Malthusians, but this has not been the case historically.

Disengaged scholars and scientists. These include geneticists, demographers, anthropologists, archeologists, sociologists, psychologists – in a word any discipline devoted entirely or in part to the study of man. This group is painfully aware of the unwritten rules of censorship with regard to qualitative studies, so that members of the scholarly and scientific community often seek refuge from ideological storms by occupying themselves with noncontroversial questions. A geneticist, for example, may devote himself to studying specific gene sequences and studiously avoid the discussion of all social implications. It is like a mechanic who repairs a carburetor with no thought as to where the automobile is to go. Some members of this particular group can be ideologized to a greater degree than nonmembers, and they can on occasion permit their personal views to influence their studies, concealing the fact not only from the public, but even from themselves. On the other hand, a large percentage remain oblivious to the philosophical and political implications of their field of study.

The Jews

Don't do what I do, do what I tell you.
Everyone's father

The popular impression is that the eugenics movement was a racist, anti-Semitic Nazi ideology inspired by Anglo-American elites. In point of fact, eugenics also managed to establish strong bridgeheads in Argentina, Australia, Austria, Belgium, Bolivia, Brazil, Canada, China, Cuba, Czechoslovakia, Denmark, Estonia, Finland, Greece, Hungary, India, Italy, Japan, Mexico, Norway, New Zealand, the Netherlands, Poland, Portugal, Rumania, Russia, South Africa, Spain, Sweden, Switzerland, and Turkey.[125]

Jews played a modest but active role in the early eugenics movement. In 1916, Rabbi Max Reichler published an article entitled "Jewish Eugenics," in which he attempted to demonstrate that Jewish religious customs were eugenic in thrust. A decade and a half later Ellsworth Huntington, in his book *Tomorrow's Children*, which was published in conjunction with the directors of the American Eugenics Society, echoed Reichler's arguments, praising the Jews as being of uniquely superior stock and explaining their achievements by a systematic adherence to the basic principles of Jewish religious law, which he also viewed as being fundamentally eugenic in nature.[126]

In the Weimar Republic many Jewish socialists actively campaigned for eugenics, using the Socialist newspaper *Vorwärts* as their chief tribune.[127] Max Levien, head of the first Munich Soviet, and Julius Moses, a member of the German Socialist Party, believed strongly in eugenics. A partial list of prominent German-Jewish eugenicists would include the geneticists Richard Goldschmidt, Heinrich Poll, and Curt Stern, the statistician Wilhelm Weinberg (coauthor of the Hardy-Weinberg Law), the mathematician Felix Bernstein, and the physicians Alfred Blaschko, Benno Chajes, Magnus Hirschfeld, Georg Löwenstein, Max Marcuse, Max Hirsch, and Albert Moll.[128] The German League for Improvement of the People and the Study of Heredity was even attacked by the

Nazi publisher Julius F. Lehmann as targeted subversion on the part of Berlin Jews.[129] Löwenstein was a member of an underground resisting the National Socialist government, and Chajes, Goldschmidt, Hirschfeld, and Poll emigrated.

In America, when the revolutionary anarchist editor of the *American Journal of Eugenics*, Moses Harman, died in 1910, Emma Goldman's magazine *Mother Earth* took over distribution. In 1933, the eugenicist and University of California professor of zoology Samuel Jackson Holmes noted the significant number of Jews in the eugenics movement and praised their "native endowment of brains," while at the same time lamenting the racial bias suffered by the Jews, which caused many of their intellectuals to be wary of non-egalitarian worldviews.[130] The American Eugenics Society itself counted Rabbi Louis Mann as one of its directors, in 1935.

One of the most prominent eugenicists was the American Herman Muller, whose mother was Jewish and who received the Nobel Prize in medicine, in 1946, for his work on genetic mutation rates. A communist, Muller spent 1933-1937 as a senior geneticist at the University of Moscow, when he wrote a letter to Stalin proposing that the Soviet Union adopt eugenics as an official policy. It was the eve of the Great Purges, and Stalin definitely disapproved of the idea, at which point Muller judged it wisest to leave for Scotland and then returned to the United States. It was in the middle of his Moscow sojourn that Muller's eugenics treatise *Out of the Night* appeared in the United States. In 1932, Muller had spent a year in Germany and he was outraged by Nazi concepts and policies concerning race.

According to the National Library in Jerusalem, from the 1920s through the 1950s, some 200 Hebrew-language Parents' manuals were published. These publications contained a coherent worldview, of which eugenics formed an integral part, subjecting Jewish mothers to an unremitting program of education, indoctrination and regulation. During the British mandate, Jewish physicians in Palestine actively promoted eugenics. Dr. Joseph Meir, for whom the hospital in Kfar Sava is named, wrote in 1934:

> *Who should be allowed to raise children? Seeking the right answer to this question, eugenics is the science that tries to refine the human race and keep it from decaying. This science is still young, but it has enormous advantages.... Is it not our duty to insure that our children will be healthy, both physically and mentally? For us, eugenics in general, and mainly the careful prevention of hereditary illnesses, has a much higher value than in other nations. Doctors, athletes, and politicians should spread the idea widely: Do not have children unless you are sure that they will be healthy, both mentally and physically.*[131]

One researcher at Ben-Gurion University working on the topic "eugenicist Zionists," came across a card file with notes written by the editors of a collection of Meir's writings, published in Israel in the mid-1950s where the editors call the article "problematic and dangerous" and comment that "Now, after Nazi eugenics, it is dangerous to publish this article."[132] In point of fact, knowledge of Jewish support for eugenics in pre-1948 Palestine was suppressed for many years.[133]

Dr. Max Nordau, the son of an Orthodox rabbi, was converted to Zionism by Theodore Herzl and became prominent in the movement. Nordau's ideas, which including vigorously propagandizing eugenics, became so popular in the Jewish community that Nordau Clubs were created even in the United States.

Dr. Arthur Ruppin, the head of the World Zionist Organization office in Palestine, wrote in his book *The Sociology of the Jews* that "in order to preserve the purity of our race, such Jews [showing signs of genetic defects] must refrain from having children."[134]

In Israel today many eugenic practices have become widely accepted. According to Meira Weiss of the Hebrew University of Jerusalem,

> *In Israel, the Zionists' eugenics turned into a selective prenatal policy backed by state-of-the-art genetic technology.*[135]

There are now more fertility clinics per capita there than in any other country in the world (four times the number per capita in the United States). Abortion is subsidized if the fetus is suspected to be physically or mentally malformed.[136]

In cases where the husband's sperm is not viable, donors fill out extensive health histories. The State supplies the sperm, which is screened for Tay-Sachs. Women over thirty-five routinely consent to amniocentesis tests and abort if genetic defects are discovered. Thus, the government is actively pursuing eugenics, although the chief motivation appears to be as least as much quantitative as qualitative.

Surrogacy was legalized in 1996[137], but only for married women. It too is paid for by the State. Jewish religious law does not delegitimize the children of unmarried women, thus making it possible to combine Jewish legal principles with modern legal practices. *In vitro* fertilization and embryo transfer are preferred by some rabbis as a form of fertility treatment that does not violate the literal Halakhic precepts against adultery[138].

Curiously, some rabbis refuse to condemn the use of non-Jewish sperm, since masturbation by non-Jews is not of explicit rabbinic concern, and also because Jewishness is passed exclusively through the mother. Children born to different Jewish mothers using the same sperm donor may even marry, since "they share no substance." Other rabbis, however, consider the use of non-Jewish sperm an abomination.[139]

The Israeli attitude toward cloning differs considerably from that prevalent in most other countries. Although human reproductive cloning is currently not permitted because the technology is not yet considered safe, the Chief Rabbinate of Israel sees no inherent religious interdiction in reproductive cloning as a form of treatment for infertility and even sees an advantage over sperm donation, which by using anonymous donors might subsequently lead to a marriage between brother and sister.[140]

In 1998, although more than eight decades had passed since the appearance of Reichler's 1916 essay, Noam J. Zohar, a professor of philosophy at Bar-Ilan University in Israel, responded to Reichler. Noting that Reichler's emphatically pro-eugenics views were "shared... by more than a few Judaic circles today," Zohar wrote that

> *A program of individualized eugenics... would seem to be consonant with an attitude that was, at the very least, tacitly endorsed by traditional Judaic teachings. Should it make a difference if the means for producing fine offspring are no longer determined by moralized speculation but instead by evidence-based genetic science?*

It seems to me that, insofar as the goal itself is acceptable, the change in the means for its advancement need pose no obstacle to its pursuit. This is so of course provided that the new means are not morally objectionable. To work out a Judaic response to the sort of new eugenics now looming on our horizon it will be necessary to evaluate the various specific means that might serve a modern individualized eugenics. I hope that some of the groundwork for that has been laid in this examination of traditional Judaic voices.[141]

The Suppression of Eugenics

Democracy demands that all of its citizens begin the race even.
Egalitarianism insists that they all finish even.
Roger Price, "The Great Roob Revolution"

Although the attack on eugenics had been launched in the late 1920s,[142] eugenics survived even the embrace of Nazi Germany, and in 1963 the Ciba Foundation convened a conference in London under the title "Man and His Future," at which three distinguished biologists and Nobel Prize laureates (Herman Muller, Joshua Lederberg, and Francis Crick) all spoke strongly in its favor. Despite this upbeat note, eugenics was about to undergo a total rout.

Outraged by pictures of police dogs attacking civil rights protesters in the South, the public found discussions of genetic racial differences intolerable. In 1974, a large group of black students descended upon the office of Professor Sandra Scarr in the Institute of Child Development of the University of Minnesota:

> *One graduate student in education said he was going to kill us if we continued to do research on black children. Another paced up and down in front of us calling, "honkie, honkie, honkie."*

When Arthur Jensen of the University of California at Berkeley visited the Institute in 1976, he and Scarr were spat upon by a phalanx of radical students, some of whom physically attacked the speakers and those who had invited him. Not only were Jensen's lectures regularly broken up, he also received bomb threats, and he had to be put under constant guard.[143]

In March 1977, the National Academy of Sciences sponsored a forum in Washington, D.C., on research with recombinant DNA. As the first session began, protestors began marching down the aisles waving placards and charts.[144]

Hans Eysenck at a lecture to have been delivered at the London School of Economics was first prevented from speaking by the chanting of "No Free Speech for Fascists!" and then physically attacked and had to be rescued from the stage, his eyeglasses broken and blood streaming from his face. When his book *The IQ Argument* appeared in the United States, wholesalers and booksellers were threatened with arson and violence, and the book became almost impossible to obtain.[145]

The above scenes, and many others like them, were triggered by assertions of mean IQs differing between racial groups, specifically between whites and blacks. No one seemed to notice that the issue was essentially irrelevant to the cause of a universalist eugenics advocated for all groups, without exception.

The second chief factor in the suppression of eugenics was the launching of the Holocaust memorial movement subsequent to the 1967 Arab/Israeli war. So effective was the campaign that polls show that many more Americans can identify the Holocaust than Pearl Harbor or the atomic bombing of Japan.[146] Those who are familiar with the term "eugenics" now associate it with "Holocaust" and "racism." The general public is totally unaware that on September 16, 1939, the leaders of the eugenics movement in the United States and England explicitly rejected the racist doctrines of the Nazi government (see Appendix 1), as did many German eugenicists. An enormous, albeit fully understandable, confusion has taken place within the Jewish community, and this confusion is fraught with significance for Jews today. According to the *National Jewish Population Survey,* Jews in America entered into a precipitous decline in numbers in the decade 1990-2000, reflecting a pattern typical of high-IQ groups.[147] Half of Jewish women aged 30-34 have no children, and nearly half of American Jews are 45 or older.[148] This is literally a matter of survival.

Beginning in the early 1980s, publications on eugenics enjoyed a considerable upswing, including a huge number of articles in the published literature and later over the Internet, but even so the majority of these publications are still either hostile or, at best, guarded. One relatively recent example is William H. Tucker's *The Science and Politics of Racial Research* (1994). While claiming to support freedom of scientific inquiry, Tucker dismisses "the trivial scientific value of IQ heritabilities," maintains that scientific rights of research "might be qualified by the rights of others," muses whether certain research topics should be pursued at all, advocates denying government funding to racial research, proposes applying the Nuremburg Code to researchers, states that the subjects of psychological research "can be wronged without being harmed" and that they should be informed of the nature of the research in case they find the results of the research unflattering. He goes on to quote the phrases "those miserable 15 IQ points" and "Are you using such gifts as you possess for or against the people?"[149] Tucker can best be seen as a moderate in the egalitarian camp.

Missa and Susanne's 1999 book *De l'eugénisme d'État à l'eugénisme privé* (From State Eugenics to Private Eugenics) is a collection of articles authored by a group of Belgian and French scholars and scientists, some of whom are hostile to eugenics while others are actually supportive. Even so, eugenics in various places is described as "utopian" and "unrealistic." Its goals are "unachievable," and it represents "a collection of false ideas" which are "contradictory" and "disproven by research." The very mention of the term can call up "unconditional condemnation for a shameful practice." Other phrases include "opprobrium," "the horrors of classical eugenics," "the danger of a eugenic drift," "American charlatans," "a dangerous trend," "the threat of eugenics," "fear," "risk," "menace," "peril," "insidious," "rampant," "radical," "immoral," "elitist," "the demon of eugenics," "the temptation of eugenics," "the worrisome Trojan horse of eugenics," "the specter of eugenics," "Nazi atrocities," "gas chambers," "racism," "ethnic discrimination," "the slippery slope of eugenics," "detestable reputation," "barbaric," "fear," "warning," "fatal," "vigilant resistance to this tendency," "genetic discrimination," "sterilizations and lobotomies," "creeping determinism," "genetic reductionism," "reduces culture to nature," "the cult of the body," "totalitarian," "utilitarian drift," "inhumane," "a mad idea," "materialist reductionism," "biologism," "geneticism," "existential or metaphysical horror," "vehement, categorical, and definitive condemnation," "universal and absolute condemnation," "absolutely evil," "worse than murder," "Thou shalt not clone!," "radical evil," "absolutely bad, absolutely contrary to good," "perversion," "intrinsically evil," "intrinsically and necessarily negative with regard to the autonomy of others," "instrumentalization and objectivization of others," "the genetic impoverishment of cloning."[150]

The campaign has been remarkably effective in achieving its goals. In 1969, *Eugenics Quarterly*, successor to *Eugenic News*, was renamed the *Annals of Human Genetics*. The following year, shortly after the first isolation of a DNA fragment which constituted a single identifiable gene, the young scientists involved in the project decided they would not continue their work on DNA. The reason, they reported, was that such work would eventually be put to evil uses by the large corporations and governments that control science.[151] Borrowing a phrase from the Soviet purges, egalitarians denounced eugenics as a "pseudo-science," so that the American Eugenics Society was forced to change its name, in 1973, to the Society for the Study of Social Biology. In 1990, the College Board changed the name of the SAT from Scholastic Aptitude Test to Scholastic Assessment Test. In 1996, it dropped the words altogether and declared that the initials no longer stood for anything whatsoever. The eugenicists themselves all ran for cover, reclassifying themselves as "population scientists," "human geneticists," "anthropologists," "demographers," and "genetic counselors."

Possible Abuse of Genetics

I am myself indifferent honest;
but yet I could accuse me of such things
that it were better my mother had not borne me.
Hamlet

Ultimately, the most serious argument militating against eugenics is its possible abuse. Unquestionably, the danger is real. It would not take much work to come up with a lengthy list of past abuses. The baby can always be drowned in the bath water. We as a species have much in our past for which we can now experience only shame.

We are just now deciphering the blueprints according to which we ourselves were constructed; we could make terrible mistakes. Or we could lose too much diversity. And as not very distant history teaches us, eugenics could be misused to justify the elimination of peoples judged "inferior" or simply hated for whatever reason. For that matter, who can possibly predict what new evils the fertile human brain is capable of in some unknown future? It is indeed frightening. Sophisticated egalitarians, who are not really egalitarians at all but simply concerned thinkers who fear the man in the street most of all, are right to experience misgivings.

The potential abuse of genetics is not limited to distorting the human genome. It is already possible to begin modifying animals to enhance their intelligence to allow them to perform tasks currently performed by people, or even to create animal-human hybrids.[152] A ready market will always exist for cheap, low-skilled workers, so that this is a real danger. Currently people feel they have the right to regard their fellow travelers on this planet as objects of consumption, so that there is not even a discussion of this frightening prospect. But imagine the moral dilemma that would face us had to deal with animals whose abilities overlapped the lower range of the human population.

Euthanasia

There is a close relationship between eugenics and the right-to-die movement. Both are philosophies of life which place value on the quality of life, not just on life *per se*.

Whereas life expectancy in England lagged behind fecundity until about 1830,[153] the average life span in modern industrial economies now extends decades beyond the fertility span. A simple visit to a nursing home provides convincing proof that there is a huge population (about to double, thanks to the baby boomers) of helpless, despairing elderly who are literally undergoing torture, day after day, month after month, year after year. Anyone who denies this obvious fact has only to change places with them – not for years, but for a few hours – to realize the tragic reality of the situation of many of them.

As we entered the third millennium, the most popular way chosen by these victims to escape their torture was to blow their brains out – a path considerably more popular among elderly men (27.7 per 100,000) than women (1.9 per 100,000).[154]

Religion

Take note, theologians, that in your desire to make matters of faith out of propositions relating to the fixity of Sun and Earth you run the risk of eventually having to condemn as heretics those who would declare the Earth to stand still and the Sun to change position.
Galileo, "The Dialogue"

There are eugenicists who believe in God, eugenicists who are agnostic, and eugenicists who are atheists. Religious belief claims to operate in a different dimension than does eugenics, although there have always been those who viewed knowledge as a replacement for religion. The Russian language, for example, amalgamates the intellectual and spiritual under a single term: *dukhovnyi*.

In one crucial aspect, however, the scientific study of human psychology is antithetical to religion. No matter what their ideologies or methods, scientists are all in hot pursuit of the holy grail of causality. This is, after all, what science is all about.

Population Management

There are two basic views of humankind: a) that we have been created in the image of God and thus are so perfect that any improvement is unthinkable; and b) that, while our species possesses great positive features as well as negative, enhancement is essential, and – at the very least – prevention of genetic decline is an absolute moral imperative.

In many ways eugenics prescribes for humankind the same goals as for non-human species: a healthy population probably limited in size so as not to upset nature's intricate balance of species and environment. Nevertheless, the specifics of human population administration are not identical either in goals or methodology to non-human population management techniques. A "drain the pond and restock" methodology is not only morally objectionable with regard to people, its feasibility is also questionable. Blatantly coercive measures can even be counter-productive when they engender resistance to eugenic reform. For eugenics as a movement to escape the temptation of utopian fantasy, it must be oriented toward the realistically achievable.

In dealing with non-domesticated animal populations, simple viability is the goal, health being defined as the capability to survive and reproduce within an environment. By contrast, human health criteria also include intelligence and altruism. As for methodology, only relative minor impingements on the wellbeing of the current human population can be tolerated, since it they and only they who can implement eugenic reform. For example, whereas wildlife managers take for granted that a balance between prey and predators is a "healthy" thing, no such Spencerian "survival of the fittest" is appropriate for humans. Despite the grand continuity of belief retained by modern eugenics from the earlier tradition, on this point realistic modern eugenics departs radically from that preached a hundred years ago.

Although individual eugenic efforts are already in full swing, they are submerged in the great demographic currents, and thus global eugenic reform is a task for society as a whole. The strength of the government relative to that of the governed population determines the limits to governmental intervention (and abuse). The weaker the government, the smaller the potential for rational population management. There is also a role to be played by non-governmental organizations, whose freedom can be less fettered than that of governments.

History is replete with instances of forced population management, the most infamous method of which is genocide. But other compulsory methods have also been employed. For example, the government of Indira Ghandi implemented a policy of compulsory sterilizations and vasectomies. And, although India ultimately came to reject this policy, the nation's current population is many millions smaller than it would have been without it. Nevertheless, China's semi-compulsory one-child policy has proven far more efficacious, and India with a Total Fertility Rate of 3.1 will soon surpass China (TFR: 1.7) as the world's most populous nation. It is estimated that by 2000 the Chinese population was already a quarter billion less than it would have been without the one-child policy. On the other hand, there are situations where emergency methods may well present the only means of averting major catastrophe. Bangladesh and Haiti come to mind, but the political will even to raise the topic is totally absent. Global society is living a fatal lie.

Shifting our focus from quantitative to qualitative questions, the debate over voluntary versus compulsory methods has thus far amounted largely to pandering to the whims of current generations. Indeed, the very phrase "reproductive rights" itself represents a bias. Do people have the "right" to give birth to babies who in all probability will grow up feeble minded or who are likely to suffer from devastating genetic illnesses? On the one side of the equation may be a single person with a genetic IQ so low that simply coping in society is well nigh impossible and, on the other, the millions of disadvantaged offspring whom he and/or she may ultimately engender over the generations. Forced sterilizations of persons with genetically predetermined low IQ and major genetic illnesses should be reinstituted. This is an unpopular statement, but it has to be said. Our current refusal to take into account the right of future generations to health and intelligence is a cowardly betrayal of our own children. Can it be that we are so selfish as to want to breed a genetically disadvantaged class of servants to perform our menial tasks for us?

The grand demographic trend is toward below-replacement fertility rates, and while compulsion has its place, the good news is that energetic voluntary measures ought usually to be sufficient to permit women of reproductive age to realize their goal of smaller families. Clearly, voluntary methods are generally preferable to compulsory, although the line between voluntarism and coercion can often be vague.

One voluntary method involves the use of ultrasound to determine the sex of the fetus. In developing countries the desire for a male offspring is often strong enough to induce parents to abort females. Ultimately the number of males in a population is reproductively insignificant, since only females can bear children, and a tiny male population is capable of impregnating a huge female population. Thus, population management has to be female-oriented.

The Chinese infant sex ratio was normal in the 1960s and 1970s (roughly 106 boys for every 100 girls), but when the one-child policy was introduced in the 1980s, the figure became far more skewed in favor of boys; by 2002 China's fifth national census revealed a sex ratio at birth of approximately 116.86 males per 100 females, having increased to 108.5 in 1982 and 110.9 in 1987. (Admittedly, there is also a question of underreporting of female births on the part of couples eager to receive permission to have another child in the hope that it will be a son.) As early as 2000 the number of men in China was already estimated to exceed that of women by sixty million.

The situation is much the same in India, where the 1991 census indicated approximately 35-45 million missing women, when ultrasound was far less available than it is now. In a ten-year study of babies born in Delhi hospitals in the period 1993-2003, the number of female births was 542 per 1,000 boys if the first child was a girl. If the first two children were girls, the ratio was only 219-1,000.

Unfortunately, although the desire for sons is greatest among rural populations, high-IQ families possess greater access to modern medicine, including ultrasound, so that this practice appears to have been dysgenic thus far. But made easily available to low-IQ families, or if such families were even financially rewarded, it could become strongly eugenic in nature, simultaneously attacking both quantitative and qualitative demographic problems. (The historic link between eugenics and Malthusian thought should be emphasized.) A sea change is already underway; by 2005 many clinics offered ultrasound for as little as 500 rupees ($11.50). It goes without saying that this is a tragic turn of events for those men who do not find a mate for themselves, but it is a far lesser evil than dysgenic overpopulation. Moreover, heightened competition for females would disproportionately reward high-IQ males. (For this same reason polygamy should be universally decriminalized. The legal enforcement of monogamy is a dysgenic intrusion into personal freedom. No scientific breeder would even consider it.)

Another voluntary method is a vigorous promotion of contraceptive methods among low-IQ families. While education is not about to cancel out the sex drive of young people, it can go a long way toward reducing the birth rate. Reversible sterilization should be actively promoted.

The current debate between "pro-choice" and "pro-life" fails utterly to take into account the consequences of abortion for genetic selection. Abortion should be actively promoted, since it often serves as the last and even only resort for many low-IQ mothers who fail to practice contraception.

Welfare policies need to be radically reexamined. Rather than simply pay low-IQ women more for each child, financial support should be made dependent on consent to undergo sterilization. Society should put more emphasis on greater tax credits for families with children, nurseries, day-care centers, etc. This would promote fertility among high-IQ women, who otherwise are tempted either not to have children at all, or to have too few, sacrificing their unborn children before the altar of career advancement. The goals of the feminist movement are in and of themselves legitimate and fair, but wed to the anti-scientific worldview of radical egalitarianism, they will devastate our species.

Eugenic family planning services are the greatest gift that the advanced countries can offer the Third World. In a global society, parochial fixation on any one country is a pathology that human society can ill afford. What is needed is tough love. Such a policy would promote the interests of any ethnic group, all of which suffer when their least intelligent members serve as the breeding pool while the most intelligent encounter strong disincentives to fertility.

In different countries a different mix of governmental and non-governmental activism is appropriate. Useful measures would include paying low-IQ women to accept embryo transfer. Sperm banks need to be encouraged to attach the greatest importance to intelligence, and the promotion of these institutions should be covered out of tax monies. And the technology should be developed to create an artificial womb or, alternatively, make inter-species embryo transplants a reality, rapidly increasing the number of high-IQ individuals.

Religious belief will always be with us, and eugenics must not be presented as scientific in an anti-religious sense. At the same time there is a huge potential for excess if eugenics were to become a core belief of the masses.

Genetic research needs to be promoted without regard to cost. Who can say what enormous potential awaits us in the future as a result of germ-line intervention?

On the immigration front, the importation of low-IQ groups to perform unskilled labor at low wages must be recognized as a threat to the host population's long-term viability. Panmixia also represents a loss in genetic diversity. All populations represent unique entities, and the loss of such uniqueness is everyone's loss. Nevertheless, given the realities of improved transportation and communication, inbreeding can only increase in the future.

Feasibility

Nature has packed away this long brain
Like a sword into scabbard.
She has forgotten those whose grave is green,
Whose breath is red, whose laugh is supple.
Osip Mandelstam, "Lamarck"

When an ideal is recognized as unachievable, it is dismissed as "utopian." If real sacrifice is required on the part of the currently living, whose altruism extends downward for only a generation or two and who for the most part are indifferent to culture and civilization, is eugenics not simply a fantasy?

To evaluate the feasibility of reestablishing the eugenics movement as a viable social force, we must first take a hard look at political systems and move beyond the populist jingoism which is as eternal as it is ubiquitous. In a dictatorship, power is patently invested in one person, whereas in "democracies" the pyramidal power structure is more opaque:

Level A: lobbies and (largely anonymous) oligarchs.

Level B: politicians.

Level C: prominent government staffers and media.

Level D: the general population.

What is crucial in this scheme of things is that the relationship of Levels B and C to Level A is, to a significant degree, that of employee to employer. To be elected, politicians need money for polling and advertising/propaganda, while the media (also owned by Level A) entertain the general population with competitions in which the differences between the competitors is minimal. Once "elected," politicians then implement the will of those who provided the financing, while losing politicians are "parked" in profitable ceremonial positions to ready themselves for the next round. To be sure, there are sophisticates within the general population who are not duped as to the nature of the system, but they can be intimidated, co-opted, or even permitted to voice discontent. Since they pose no threat to the system, their protests are used as a demonstration of "freedom of speech." The bottom line is that all human social structures are oligarchic in nature, and the implementation of a viable eugenics policy is dependent on a relatively tiny elite.

Eugenics is not an either/or proposition. Many of the decisions being taken on a governmental level are already fraught with genetic consequences – family planning programs, legalized and subsidized abortions, immigration criteria, tax credits for having children, mandated paid parental leave, genetic research, cloning, fertility assistance, and so on. Eugenicists argue that it is only reasonable that the decision makers take into account the eugenic or dysgenic consequences of governmental actions.

The world is divided into independent nations. Given the necessary funding, it would be possible in at least some of them to set up positive-eugenic breeding programs which would not necessarily depend on human birth mothers. The resistance to such changes is understandably intense, considering that even artificial insemination continues to be resisted in some quarters.

One obvious factor that will promote the eugenic agenda is the undeniable desire of parents to have healthy, intelligent children. Genetic screening of embryos will obviously encompass a greater and greater range of detectable traits, and thus the bar will be raised from simply eliminating disastrous diseases to attempting to produce children who enjoy genetic advantages that are currently available to a smaller percentage of the population. Germ-line therapy, unlike both the traditional methods of positive and negative eugenics, will make it possible for people to have their own children – but children who will be more healthy and intelligent than they would have been without genetic intervention. This method will entirely bypass the intergenerational conflict of interests which works to the disadvantage of the helpless unborn.

As discussed above, public opinion is extremely malleable. Advertising and political propaganda come down to cost. But if any individual country were to aggressively pursue a national eugenics policy while being militarily weak, of if any ethnic group were to follow such a course of action, non-participating countries/groups would sense a competitive threat to their offspring and would be sorely tempted to launch a preemptive strike so as to avoid the necessity of introducing a eugenics policy themselves.

Radical Intervention

We know what we are, but not what we may be.
Hamlet

While we are still at an extremely early stage in our understanding of human genetics, it is entirely foreseeable that future knowledge will permit us to go beyond simple genetic tinkering to replace this or that disease-engendering gene or enhance some desirable ability or personality trait. We will be able to go much further and alter the genetic constitution in the most radical fashion. As pointed out by the bioethicist and theologian Joseph Fletcher as early as 1973, the creation of persons whose genome is partly borrowed from other species is entirely possible.[155] Recent writing now discusses the "fungibility" of DNA, the consequent malleability of life, the fact that human nature is not fixed, the possibility that at some future point different groups of human beings may follow divergent paths of development through the use of genetic technology – perhaps as different from one another as men and women are now, the collapse of interspecies barriers, the possibility of not simply discovering genes but creating them. Should we really attempt to preserve human nature or should we attempt to change it?[156]

John H. Campbell, a biologist at the University of California, is among those who advocate radical interventionism. He writes that

> *Geneticists are laying open our heredity like the circuit board of a radio.... We shall be able to redesign our biological selves at will.... In point of fact, it is hard to imagine how a system of inheritance could be more ideal for engineering than ours is.*[157]

Reasoning that the majority of humankind will not voluntarily accept qualitative population-management policies, Campbell points out that any attempt to raise the IQ of the whole human race would be tediously slow. He further points out that the general thrust of early eugenics was not so much species improvement as the prevention of decline.

Campbell's eugenics, therefore, advocates the abandonment of *Homo sapiens* as a "relic" or "living fossil" and the application of genetic technologies to intrude upon the genome, probably writing novel genes from scratch using a DNA synthesizer. Such eugenics would be practiced by elite groups, whose achievements would so quickly and radically outdistance the usual tempo of evolution that within ten generation the new groups will have advanced beyond our current form to the same degree that we transcend apes.

Campbell anticipates the creation of new species according to the punctuated equilibrium scenario discussed earlier. Practitioners of the new eugenics would view themselves as intermediaries of evolution rather than as finished products. Freed from the "drag" of an outdated species that is already in decline, they could evolve in intelligence in a geometrical increase – forever. Our current intellect, Campbell projects, is probably unable even to comprehend the mental attributes that descendants will struggle to conceive. He then goes on to advocate an old idea – eugenic religions. Not accidentally, one of the sites circulating Campbell's article is that of "Prometheism." Lastly, he points out that some appropriate genetic technologies are already available:

> *Private autoevolution is not a possibility for a distant future nor is it a science fiction. It is with us now, albeit at an early enough phase to have escaped most people's attention.... The most significant legacy of our age will not be nuclear power, computers, political achievements or a static ethics for a "sustainable" society. It will be the closure of our rational intellect around our evolution. The statues of the 21st century will celebrate the fathers of* Homo autocatalyticus *who brought evolution under its own reason. The world waits to see whose faces will adorn them.* [158]

Campbell's projection of rapid, small-group-directed evolution is at once heartening and depressing. Greater, even open-ended, intelligence is awesome to contemplate. On the other hand, how sad it is for those "living fossils" who constitute the mass of humanity – humanity, at least, as we know it today.

The reader will recall that eugenics does not limit itself to the present population but defines society as the entire human community over time; the movement perceives itself as the fourth leg of the table upon which that community rests. (The three other legs are a supply of natural resources; a clean, biodiverse environment; and a human population no larger than the planet can comfortably sustain on an indefinite basis.) This means that we are dealing with what eugenicists consider to be non-negotiable issues. Such conditions are viewed as either essential to survival or intrinsically linked to the very meaning of existence. All other considerations – political parties, for example, or even the welfare of today's population – are perceived as flowing from and subordinate to these fundamental necessities.

What this means is that if the eugenics platform is to have any chance of success it will have to adopt a posture of non-partisanship and link itself to neither the political right nor the left. At the same time, for strategic considerations, the movement cannot afford embroilment in inter-group conflict or even inter-group comparisons. While these areas may constitute legitimate concerns for the political scientist, the sociologist, or the human biologist, history has demonstrated that their pursuit within the eugenic agenda can be counterproductive and even disastrous. Scholars and scientists wishing to promote the eugenics agenda will have to search for commonalities with other thinkers rather than enter into conflict with them. Ideological separation will require a self-discipline that no one will readily embrace. To be honest, some of these topics can be of eugenic significance. At the very least, they can intersect with eugenic considerations.

Presently, such self-control is not even being attempted. A post-human or even a non-human evolutionary path to intelligence – as opposed to a general uplifting of the whole population – therefore appears more and more likely.

Legal barriers are already being erected in a frantic attempt to prevent a resurgence of eugenics, but to believe that such measures can be completely effective is a hopeless fantasy. Campbell's logic is inescapable. The rejection of traditional within-species eugenics – despite all the posturing of society – will inevitably lead to the scenario he describes.

The invention of writing created a global human mind, in which knowledge is passed on and accumulated over generations. In the process, individual people specialize in specific fields, and no one today would be tempted to speak of "universal geniuses." There is simply too much to know.

While the human brain has been millions of years in the making, computers, which have been in development really for only about a century, are already beating the best human players at chess. "Hal" may not yet have been born but he is even now kicking in his binary womb.

Carbon-based technology has its limitations. The individual human brain is limited by its size, by the amount of time available for learning, and by the speed at which it can process information. A computer can be created of any size with limitless memory and limitless programming. As for speed, current technology is already processing information in picoseconds (trillionths of a second), whereas the human brain is capable of mere microseconds.[159]

The human mind is itself a machine, and its quirks, self-consciousness, and adaptability will all eventually be explained, even though we are only beginning to unlock its secrets. Currently a noisy debate is ongoing as to whether computer brain power can surpass human, but really it is a question of when rather than whether. The two societies projected by H.G. Wells in *The Time Machine*, one producing material goods and the other, childlike, consuming them, is probably going to arrive sooner than we think, and the childlike creatures will be us.

This soon-to-be reality relegates to eugenics a far more modest role than would otherwise be imaginable. Any effort to improve the human brain is targeted at an instrument which is inherently limited in its capacity. The machine brain, on the other hand, will be something like God.

Allotted only a thousand months or so of existence, we individuals are as ephemeral as chaff in the wind, but the fate of thought, of culture, of life itself has been thrust upon us, and we can either fritter away the patrimony of millions of generations in the gratification of individualistic and tribal instincts or we can stride forward to fulfill our fate, shouldering our responsibilities to a future world and linking hands in the great chain of generations.

Conclusion

A father's responsibility
Deuteronomy 6:1-9

As the collective human brain ponders both its own origins and its future, the eugenics platform reemerges as timeless, for the issues it deals with are independent of both historical advocacy and repudiation by individuals.

The left-right political continuum has been set according to issues of importance to currently living constituencies, whose interests are largely peripheral and even instrumental to the eugenics platform, where neither the expanded (longitudinal) definition of humankind nor the teleology of existence fit into the accepted spectrum.

The conflict of interests between us and future generations represents a moral confrontation, but politics can best be summarized as the formation of alliances based on mutual advantage. Which are the constituencies that will agree to partner with future generations when no *quid pro quo* is possible? Do such constituencies even exist?

What You Can Do For Future Generations

1. Tell your friends about this book and forward to them the website at which the book can be downloaded free of charge: www.whatwemaybe.org.

2. If you are a native speaker of a language other than English and wish to volunteer to translate this book into your native tongue, please contact Dr. Glad. Dr. Glad's current e-mail address my be learned from the electronic version of the text, available at www.whatwemaybe.org.

3. Assign the book to your students if you are a teacher dealing with any of the following areas: academic freedom, anthropology, bioethics, biology, biopolitics, cloning, crime, demographics, ecology, egalitarianism, environmentalism, ethics, eugenics, euthanasia, evolution, fertility, futurology, generational equity, genetics, history, the holocaust, human rights, migration / emigration / immigration), philosophy, political science, population studies, religion, sociobiology, sociology, testing, welfare.

Appendix 1
Social Biology and Population Improvement

The following document, which appeared in Nature, September 16, 1939, was a joint statement issued by America's and Britain's most prominent biologists (some of them Nobel Prize laureates), and was widely referred to as the "Eugenics Manifesto." The Second World War had already begun, and the authors explicitly decried antagonism between races and theories according to which certain good or bad genes are the monopoly of certain peoples. The document is published here in its entirety.

Social Biology and Population Improvement

In response to a request from Science Service, of Washington, D.C., for a reply to the question "How could the world's population be improved most effectively genetically?", addressed to a number of scientific workers, the subjoined statement was prepared, and signed by those whose names appear at the end.

The question "How could the world's population be improved most effectively genetically?" raises far broader problems than the purely biological ones, problems which the biologist unavoidably encounters as soon as he tries to get the principles of his own special field put into practice. For the effective genetic improvement of mankind is dependent upon major changes in social conditions, and correlative changes in human attitudes. In the first place, there can be no valid basis for estimating and comparing the intrinsic worth of different individuals, without economic and social conditions which provide approximately equal opportunities for all members of society instead of stratifying them from birth into classes with widely different privileges.

The second major hindrance to genetic improvement lies in the economic and political conditions which foster antagonism between different peoples, nations and 'races'. The removal of race prejudices and of the unscientific doctrine that good or bad genes are the monopoly of particular peoples or of persons with features of a given kind will not be possible, however, before the conditions which make for war and economic exploitation have been eliminated. This requires some effective sort of federation of the whole world, based on the common interests of all its peoples.

Thirdly, it cannot be expected that the raising of children will be influenced actively by considerations of the worth of future generations unless parents in general have a very considerable economic security and unless they are extended such adequate economic, medical, education and other aids in the bearing and rearing of each additional child that the having of more children does not overburden either of them. As the woman is more especially affected by childbearing and rearing, she must be given special protection to ensure that her reproductive duties do not interfere too greatly with her opportunities to participate in the life and work of the community at large. These objects cannot be achieved unless there is an organization of production primarily for the benefit of consumer and worker, unless the conditions of employment are adapted to the needs of parents and especially of mothers, and unless dwellings, towns and community services generally are reshaped with the good of children as one of their main objectives.

A fourth prerequisite for effective genetic improvement is the legalization, the universal dissemination, and the further development through scientific investigation, of ever more efficacious means of birth control, both negative and positive, that can be put into effect at all states of the reproductive process – as by voluntary temporary or permanent sterilization, contraception, abortion (as a third line of defence), control of fertility and of the sexual cycle, artificial insemination, etc. Along with all this the development of social consciousness and responsibility in regard to the production of children is required, and this cannot be expected to be operative unless the above-mentioned economic and social conditions for its fulfillment are present, and unless the superstitious attitude towards sex and reproduction now prevalent has been replaced by a scientific and social attitude. This will result in its being regarded as an honour and a privilege, if not a duty, for a mother, married or unmarried, for a couple, to have the best children possible, both in respect of their upbringing and of their genetic endowment, even where the latter would mean an artificial – though always voluntary – control over the process of parenthood.

Before people in general, or the State which is supposed to represent them, can be relied upon to adopt rational policies for the guidance of their reproduction, there will have to be, fifthly, a far wider spread of knowledge of biological principles and of recognition of the truth that both environment and heredity constitute dominating and inescapable complementary factors in human wellbeing, but factors both of which are under the potential control of man and admit of unlimited but interdependent progress. Betterment of environmental conditions enhances the opportunities for genetic betterment in the ways above indicated. But it must be also understood that the effect of the bettered environment is not a direct one on the germ cells and that the Lamarckian doctrine is fallacious, according to which the children of parents who have had better opportunities for physical and mental development inherit these improvements biologically, and according to which, in consequence, the dominant classes and people would have become genetically superior to the underprivileged ones. The intrinsic (genetic) characteristics of any generation can be better than those of the preceding generation only as a result of some kind of selection, that is, by those persons of the preceding generation who had a better genetic equipment have produced more offspring, on the whole, than the rest, either through conscious choice, or as an automatic result of the way in which they lived. Under modern civilized conditions such selection is far less likely to be automatic than under primitive conditions, hence some kind of conscious guidance of selection is called for to make this possible, however, the population must first appreciate the force of the above principles, and the social value which a wisely guided selection would have.

Sixthly, conscious selection requires, in addition, an agreed direction or directions for selection to take, and these directions cannot be social ones, that is, for the good of mankind at large, unless social motives predominate in society. This in turn implies its socialized organization. The most important genetic objectives, from a social point of view, are the improvement of those genetic characteristics which make (a) for health, (b) for the complex called intelligence, and (c) for those temperamental qualities which favour fellow-feeling and social behaviour rather than those (to-day most esteemed by many) which make for personal 'success', as success is usually understood at present.

A more widespread understanding of biological principles will bring with it the realization that much more than the prevention of genetic deterioration is to be sought for, and that the raising of the level of the average of the population nearly to that of the highest now existing in isolated individuals, in regard to physical wellbeing, intelligence and temperamental qualities, is an achievement that would – so far as purely genetic considerations are concerned – be physically possible with a comparatively small number of generations. Thus everyone might look upon 'genius,' combined of course with stability, as his birthright. As the course of evolution shows, this would represent no final stage at all, but only an earnest of still further progress in the future.

The effectiveness of such progress, however, would demand increasingly extensive and intensive research in human genetics and in the numerous fields of investigation correlated therewith. This would involve the co-operation of specialists in various branches of medicine, psychology, chemistry and, not least, the social sciences, with the improvement of the inner constitution of man himself as their central theme. The organization of the human body is marvelously intricate, and the study of its genetics is beset with special difficulties which require the prosecution of research in this field to be on a much vaster scale, as well as more exact and analytical, than hitherto contemplated. This can, however, come about when men's minds are turned from war and hate and the struggle for the elementary means of subsistence to larger aims, pursued in common.

The day when economic reconstruction will reach the stage where such human forces will be released is not yet, but it is the task of his generation to prepare for it, and all steps along the way will represent a gain, not only for the possibilities of the ultimate genetic improvement of man, to a degree seldom dreamed of hitherto, but at the same time, more directly, for human mastery over those more immediate evils which are so threatening our modern civilization.

Signatories: F. A. E. Crew, C. D. Darlington, J. B. S. Haldane, S. C. Harland, L. T. Hogben, J. S. Huxley, H. J. Muller, J. Needham, G. P. Child, P. R. David, G. Dahlberg, Th. Dobzhansky, R. A. Emerson, C. Gordon, J. Hammond, C. L. Huskins, P. C. Koller, W. Landauer, H. H. Plough, B. Price, J. Schultz, A. G. Steinberg, C. H. Waddington.[160]

Appendix 2
100 Books Dealing with German History during the Weimar Period and under National Socialism

Books with no references to eugenics in index

1. Abel, Theodore. 1938, 1966. *The Nazi Movement*. Atherton Press. **2.** Abel, Theodore. 1938. *Why Hitler Came into Power*. Prentice-Hall. **3.** Arendt, Hannah. 1965. *Eichmann in Jerusalem: A Report on the Banality of Evil*. Viking Press. **4.** Baird, Jay W. 1990. *To Die for Germany: Heroes in the Nazi Pantheon*. Indiana University Press. **5.** Barnouw, *DagMarch* 1988. *Weimar Intellectuals and the Threat of Modernity*. Indiana University Press. **6.** Berg-Schlosser, Dirk; Rytlewski, Ralf (eds). 1993. *Political Culture in Germany*. St. Martin's Press. **7.** Brecht, Arnold. 1944. *Prelude to Silence: The End of the German Republic*. Oxford University Press, New York. **8.** Bullock, Alan. 1962. *Hitler: A Study in Tyranny*. Harper & Row. **9.** Carsten, Francis L. 1965. *Reichswehr und Politik 1918-1933*. Kiepenheuer & Witsch. Reissued in English in 1966 by Oxford at the Clarendon Press. **10.** Cecil, Robert. 197. *The Myth of the Master Race: Alfred Rosenberg and Nazi Ideology*. Dodd Mead & Company. **11.** Childs, David. 1991. *Germany In the Twentieth Century*. HarperCollins Publishers. **12.** Compton, James V. 1967. *The Swastika and the Eagle: Hitler, the United States, and the Origins of World War II*. Houghton Mifflin Company. **13.** Goldensohn, Leon. 2004. *Nuremburg Interviews: An American Psychiatrist's Conversations with Defendants and Witnesses,* Knopf. **14.** Davidson, Eugene. 1996. *The Unmaking of Adolf Hitler*. University of Missouri Press. **15.** Diehl, James M. 1977. *Paramilitary Politics in Weimar Germany*. Indiana University Press. **16.** Dobkowski, Michael N.; Wallimann, Isidor. 1989. *Radical Perspectives on the Rise of Fascism in Germany 1919-1945*. Monthly Review Press. **17.** Eksteins, Modris. 1975. *The Limits of Reason: The German Democratic Press and the Collapse* *of Weimar Democracy.* Oxford University Press. **18.** Eschenburg, Theodor; Fraenkel, Ernst; Sontheimer, Kurt; Matthis, Erich; Morsey, Rudolph; Flechtheim, Ossip K.; Bracher, Karl Dietrich; Krausnick, Helmut; Rothfels, Hans; Kogon, Eugen. 1966. *The Path to Dictatorship 1918-1933: Ten Essays.* Frederick A. Praeger. **19.** Eyck, Erich. 196. *A History of the Weimar Republic.* Harvard. **20.** Farago, Ladislas. 1974. *Aftermath: Martin Bormann and the Fourth Reich.* Simon Schuster. **21.** Feuchtwanger, E. J. 1995. *From Weimar to Hitler: Germany 1918-1933.* St. Martin's Press. **22.** Fraser, Lindley. 1945. *Germany Between Two Wars: A Study of Propaganda and War-Guilt.*Oxford University Press. **23.** Frazer, David. 1993. *Knight's Cross: A Life of Field Marshal Erwin Rommel.* HarperCollins. **24.** Fried, Hans Ernest. 1943. *The Guilt of the German Army.* The Macmillan Company. **25.** Fritsche, Peter. 1998. *Germans Into Nazis.* Harvard University Press. **26.** Fritzsche, Peter. 1990. *Rehearsals for Fascism: Populism and Political Mobilization in Weimar Germany.* Oxford University Press. **27.** Fulbrook, Mary. 1992. *The Divided Nation: a History of Germany 1918-1990.* Oxford University Press. **28.** Guérin, Daniel. 1994. *The Brown Plague: Travels in late Weimar & Early Nazi Germany.* Duke University Press. **29.** Halperin, S. William. 1965. *Germany Tried Democracy: A Political History of the Reich from 1918 to 1933.* Norton. **30.** Hamann, Brigitte. 1999. *Hitler's Vienna: A Dictator's Apprenticeship.* Oxford University Press. **31.** Hanser, Richard. 1970. *Putsch! How Hitler Made Revolution.* Peter H. Wyden, Inc. **32.** Heiber, Helmut. 1972. *Goebbels.* Hawthorn Books. **33.** Heiber, Helmut. 1974. *Die Republik von WeiMarch* Deutscher Taschenbuch Verlag. Reissued in English in 1993 by Blackwell. **34.** Heiden, Konrad. 1944. *The Führer.* Carroll & Graf Publishers. **35.** Herzstein, Robert Edwin. 1974. *Adolf Hitler and the German Trauma 1913-1945.* Capricorn Books. **36.** Heydecker, Joe J.; Leeb, Johannes. 1962. *The Nuremberg Trial: A History of Nazi Germany As Revealed Through the Testimony at Nuremberg.* Greenwood Press. **37.** Hiden, J. W. 1974. *The Weimar Republic.* Longman. **38.** Hilger, Gustav; Meyer, Alfred G. Meyer. 1953. *The Incompatible Allies: A*

Memoir-History of German-Soviet Relations 1918-1941. Macmillan. **39.** Hitler, Adolf. 1942. *The Speeches of Adolf Hitler April 1922 – August 1939*. Oxford University Press. **40.** Hitler, Adolf. 1971. *Mein Kampf*, Houghton Mifflin Company. **41.** Homer, F. X. J.; Wilcox, Larry, D. 1986. *Germany and Europe in the Era of the Two Word Wars*, University Press of Virginia. **42.** Housden, Martyn. 2000. *Hitler: Study of a Revolutionary?* Routledge. **43.** de Hoyos, Ladislas. 1985. *Klaus Barbie*. W. H. Allen. **44.** Hughes, John Graven. 1987. *Getting Hitler into Heaven*. Corgi Books. **45.** Jablonsky, David. 1989. *The Nazi Party in Dissolution: Hitler and the Verbotzeit 1923-1925*. Frank Cass. **46.** Shirer, William L. 1990. *The Rise and Fall of the Third Reich: A History of Nazi Germany,* Touchstone Books. **47.** Jasper, Gotthard. 1968. *Von Weimar zu Hitler 1930-1933*. Kiepenheuer & Witsch. Jetzinger, Franz. 1958, 1976. *Hitler's Youth*. Greenwood Press. **48.** Jones, J. Sydney. 1983. *Hitler in Vienna 1907-1913*. Stein and Day Publishers. **49.** Jones, Nigel H. 1987. *Hitler's Heralds: The Study of the Freikorps 1918-1923*, John Murray. **50.** Kastning, Alfred. 1970. *Die deutsche Sozialdemokratie zwischen Koalition und Opposition*. Ferdinand Schöningh. **51.** Kersten, Felis (ed.: Herma Briffault). 1947. *The Memoirs of Doctor Felix Kersten*. Doubleday & Co. **52.** Kilzer, Louis. 2000. *Hitler's Traitor: Martin Bormann and the Defeat of the Reich*. Presidio. **53.** Klemperer (von), Klemens. 1957, 1968. *Germany's New Conservatism: Its History and Dilemma in the Twentieth Century*, Princeton University Press. **54.** Kochan, Lionel. 1963. *The Struggle for Germany 1914-1945*. Edinburgh at the University Press. **55.** Koch-Weser, Erich. 1930. *Germany in the Post-War World*. Dorrance & Co. **56.** Koenisberg, Richard A. 1975. *Hitler's Ideology: A Study in Psychoanalytic Sociology*. The Library of Social Science. **57.** Könneman, Erwin; Krusch, Hans-Joachim. 1972. *Aktionseinheit contra Kapp-Putsch*. Dietz Verlag. **58.** Kosok, Paul. 1933. *Modern Germany: A Study of Conflicting Loyalties*. University of Chicago Press. **59.** Langer, Walter C. *The Mind of Adolf Hitler: The Secret Wartime Report*. Basic Books. **60.** Lee, Marshall M.; Michalka, Wolfgang. 1987. *German For-*

eign Policy 1917-1933. Berg. **61.** Linklater, Magnus; Hilton, Isabel; Ascherson, Neal. 1985. *The Nazi Legacy: Klaus Barbie and the International Fascist Connection.* Holt, Rinehart and Winston. **62.** Ludecke, Kurt G. W. 1937. *I Knew Hitler.* Charles Scribners. **63.** Manvell, Roger; Fraenkl, Heinrich. 1969. *The Canaris Conspiracy: The Secret Resistance to Hitler in the German Army.* David McKay Company. **64.** McKenzie, John R. P. 1971. *Weimar Germany 1918-1933.* Rowman and Littlefield. **65.** Merker, Paul. Vol. 1, 1944, Vol. 2, 1945. *Deutschland: Sein oder nicht sein?* El Libro Libre, Mexico City. **66.** Messenger, Charles. 1991. *The Last Prussian: A Biography of Field Marshal Gerd von Rundstedt 1875-1953.* Brassey's. **67.** Mitcham, Samuel W. 1996. *Why Hitler? The Genesis of the Nazi Reich*, Praeger. **68.** Mommsen, Hans. 1991. *From Weimar to Auschwitz.* Princeton University Press. **69.** Morgan, J. H. 1945. *Assize of Arms: Being the Story of the Disarmament of Germany and Her Rearmament 1919-1939.* Methuen & Co. **70.** Murphy, David Thomas. 1997. *The Heroic Earth: Geopolotical Thought in Weimar Germany 1918-1933.* Kent State University Press. **71.** Nicholls, A. J. 1991. *Weimar and the Rise of Hitler.* St. Martin's Press. **72.** Nicholls, Anthony; Matthias, Erich (eds.). 1971. G*erman Democracy and the Triumph of Hitler.* George Allen and Unwin. **73.** Pachter, Henry. 1982. *Weimar Studies.* Columbia University Press. **74.** Paris, Erna. 1986. *Unhealed Wounds: France and the Klaus Barbie Affair.* Grove Press. **75.** Patch, William L. 1998. *Heinrich Brüning and the Dissolution of the Weimar Republic.* Cambridge University Press. **76.** Payne, Robert. 1973. *The Life and Death of Adolf Hitler.* Praeger. **77.** Peterson, Edward N. 1969. *The Limits of Hitler's Power.* Princeton University Press. **78.** Pool, James. 1997. *Hitler and His Secret Partners: Contributions, Loot and Rewards 1933-1945.* Pocket Books. **79.** Price, G. Ward. 1938. *I Know These Dictators.* Henry Holt and Company. **80.** Price, Morgan Philips. 1999. *Dispatches from the Weimar Republic: Versailles and German Fascism.* Pluto Press. **81.** Robinson, Jacob. 1965. *And the Crooked Shall Be Made Straight: The Eichmann Trial, the Jewish Catastrophe, and Hannah Arendt's Narra-*

tive. Macmillan. **82.** Roll, Erich. 1933. *Spotlight on Germany: A Survey of Her Economic and Political Problems*. Faber & Faber Limited. **83.** Russell (Lord) of Liverpool. 1963. *The Record: The Trial of Adolf Eichmann for His Crimes Against the Jewish People and Against Humanity*. Alfred A. Knopf. **84.** Schacht, Hjalmar Horace Greeley. 1974. *Confessions of "The Old Wizard": Autobiography*. Greenwood Press. **85.** Scheele, Godfrey. 1946. *The Weimar Republic: Overture to the Third Reich*. Faber and Faber Limited. **86.** Schellenberg, Walter. 1956. *The Labyrinth: Memoirs*. Harper and Brothers Publishers. **87.** Schultz, Sigrid. 1944. *Germany Will Try It Again*. Reynal & Hitchcock. **88.** Stachura, Peter D. 1983. *The Nazi Machtergreifung*. George Allen & Unwin. **89.** Stachura, Peter D. 1993. *Political Leaders in Weimar Germany: A Biographical Study*. Simon & Schuster. **90.** Taylor, Simon. 1983. *The Rise of Hitler: Revolution and Counter-Revolution in Germany 1918-1933*. Universe Books. **91.** Dederke, Karlheinz. 1984. *Reich und Republik Deutschland 1917-1933*. Klett-Cotta. **92.** Villard, Oswald Garrison. 1933. *The German Phoenix: The Story of the Republic*. Harrison Asmith & Robert Haas. **93.** Waite, Robert G. L. 1952. *Vanguard of Nazism: The Free Corps Movement in Post-War Germany*. Harvard. **94.** Watkins, Frederick Mundell. 1939. *The Failure of constitutional emergency Powers under the German Republic*. Harvard University Press. **95.** Welch, David. 1983. *Nazi Propaganda: The Power and The Limitations*. Croom Helm & Barnes & Noble Books. **96.** Wheeler-Bennett, John W. 1967. *The Nemesis of Power: The German Army in Politics 1918-1945*. Viking Press.

Books with references to eugenics in index

97. Benderesky, Joseph W. 1956. *A History of Nazi Germany*. Burnham Inc. According to the index, eugenics is mentioned on mentioned on 10 pages, but several of these actually refer to euthanasia rather than eugenics, and the others are limited to Hitler's belief in "Aryan" racial superiority. **98.** Bramwell, Anna. 1985. *Blood and Soil: Richard Walther Darré and Hitler's "Green Party,"* Kensal Press, 7 mentions. **99.** Hiden,

John. 1996. *Republican and Fascist Germany: Themes and Variations in the History of Weimar and the Third Reich 1918-1945,* Longman, 2 mentions. **100.** Peukert, Detlev J. K.1991. *The Weimar Republic: The Crisis of Classical Modernity,* Hill and Wang, 2 mentions.

Works Cited

American Association for the Advancement of Science (AAAS). 2000. "The Human Genome," *Science*, special issue, Vol. 291, No. 5507.

Associated Press. 1992. "Study Shows Brains Differ in Gay, Heterosexual Men: Anterior Commissure Area Larger in Homosexuals," *Washington Post*, August 1, A2.

Associated Press. 2001a. "Population rises halt in developed nations," *Washington Times*, May 22, A6; quoting Population Reference Bureau.

Associated Press. 2001b. "Scientist says he will clone humans in U.S. or abroad," *Washington Times*, December 15, A2.

Atkinson, Richard. 2001. "SAT Is to Admissions as Inadequate Is to..." *Washington Post*, March 26, A1.

Bailey, Michael; Pillard, Richard C. 1991. "A Genetic Study of Male Sexual Orientation," *Arch. Gen. Psychiatry*, 48, 1089-96.

Bajema, Carl Jay. 1976. *Eugenics Then and Now*. Dowden, Hutchinson & Ross, Stroudsburg, Pennsylvania.

Balter, Michael. 2001. "Anthropologists Duel Over Modern Human Origins," *Science*, March 2, Vol. 291, 1728-1729.

Baur, Erwin; Fischer, Eugen; Lenz, Fritz. 1931. Human Heredity. The Macmillan Company, New York.

Bearden, H. Joe; Fuquay, John W. 2000. *Applied Animal Reproduction* (Fifth Edition). Prentice Hall, Upper Saddle River, New Jersey.

Binding, Karl; Hoche, Alfred. 1920. *Die Freigabe der Vernichtung lebensunwerten Lebens*. F. Meiner, Leipzig.

Blank, Robert H. 1982. *Torts for Wrongful Life: Individual and Eugenic Implications*. Social Philosophy and Policy Center, Bowling Green, Ohio.

Bodart, Gaston. 1916. *Losses of Life in Modern Wars*. H. Milford, London/New York.

Borkenau, Peter; Riemann, Rainer; Agleittner, Alois; Spinath, Frank M. 2001. "Genetic and Environmental Influences on Observed Personality: Evidence from the German Observational Study of Adult Twins," *Journal of Personality and Social Psychology*, Vol. 80, No. 4, 655-668.

Bowler, Peter J. 1986. *Theories of Human Evolution: A Century of Debate, 1844-1944*. Johns Hopkins University Press, Baltimore/London.

Bravin, Jess; Regaldo, Antonio. 2003. "U.N. Puts Off Human-Clone Ban Amid Demands by U.S., Vatican," *Wall Street Journal*, November 7, A3.

Brock, Dan; Buchanan, Allen; Daniels, Norman; Wickler, Daniel. 2000. *From Chance to Choice: Genes And The Just Society*. Cambridge University Press, Cambridge, U.K./New York.

Broyde, Machael J. Undated, between 1997 and 2002. "Cloning People and Jewish Law: A Preliminary Analysis." WWW, jlaw.com/Articles/Cloning.html.

Campbell, John H. 1995. Taken from Evolution and Human Values. 1995. Campbell, J. H.; Wesson, R.; and Williams, P. (editors) Rodopi Press, Amsterdam, 79-114. www.home.comcast.net/~neoeugenics/camp.htm.

Campbell, Joseph. *The Power of Myth*. Interview with Bill Moyers, Public Television.

Cavalli-Sforza, L. L.; Bodmer, W. F. 1971. *The Genetics of Human Populations*. W. H. Freeman, San Francisco.

Central Committee of the Communist Party of the Soviet Union. 1936. ""On Pedological Distortions in the Commissariats of Education," *Pravda*, July 5.

Christians for the Cloning of Jesus. "The Shroud of Turin." www.geocities.com/Athens/Acropolis/8611/page2.htlm.

Clark, A. J. 1998. *Animal Breeding: Technology for the 21st Century*, Harwood Academic, multiple publishing sites.

Cole, Tim. 1999. *Selling the Holocaust: From Auschwitz to Schindler: How History is Bought, Packaged, and Sold*. Routledge, New York.

Collange, Jean François; Houdebine, Louis-Marie; Huriet, Claude; Lecourt, Dominique; Renard, Jean-Paul; Testart, Jacques. 1999. *Faut-il vraiment cloner l'homme?* Presses universitaires de France, Paris.

Cooperman, Alan. 2002. "Number of Jews in U.S. Falls 5 Percent: Report Cites Couples' Decision to Delay Having Children as Principal Cause, " *Washington Post*, October 9, A3.

Crew, F. A. E.; Darlington, C. D.; Haldane, J. B. S. Harland, S. C.; Hogben, L. T.; Huxley, J. S. Muller, H. J.; Needham, J.; Child, G. P.; David, P. R.; Dahlberg, G.; Dobzhansky, Th.; Emerson, R. A.; Gordon, C.; Hammond, J.; Huskins, C. L.; Koller, P. C.; Landauer, W.; *Plough, H. H.; Price, B.; Schultz, J.; Steinberg, G.; Waddington, C. H.* "Social Biology and Population Improvement, " *Nature*, Vol. 144, No. 3646, 521-522.

De Marco, Donna. 2001. "What's in a name?: For direct marketers, a gold mine of data about a consumer's tastes, pocketbook," *Washington Times*, June 17, A1, 6.

"Disability Rights Advocates," Center for Genetics and Society www.genetics-and-society.org/constituencies /disability.html.

"Docs Grow Heart Cells," 2001. DNA Diagnostics Center, August 2, www.dnacenter.com/geneticnews.html.

Domhoff, G. William. 1983. *Who Rules America Now? A View for the '80s*. Prentice Hall, Englewood Cliffs, New Jersey.

Dougherty, Carter. 2001. "Free censorship with purchase? ISP blocks access to sites without consent to curb 'spam,'" *Washington Times*, May 30, B8, 9.

Drouard, Alain. 1999. L'eugénisme en questions: L'exemple de l'eugénisme "français." Ellipses, Paris.

Duster, Troy. 1990. *Backdoor to Eugenics*. Routledge, New York/London.

Eisenberg, Daniel. 2002. "The Ethics of Cloning." www.us-israel.org /jsource /Judaism /clone /html.

Elliman, Wendy. 2001. "Statistical probabilities and probable cures," *Jerusalem Post*, February 27, WWW.

Encyclopedia Britannica. 2001. "Genetic disease, human." WWW.

"Eugenics – Euthenics – Euphenics," www.bioethicsanddisability.org/ Eugenics%20Euthenics,%20Euphenics.html.

Eysenck, H. J. 1982. "The sociology of psychological knowledge, the genetic interpretation of the IQ, and Marxist-Leninist ideology," *Bulletin of the British Psychological Society*, No. 35, 449-451.

Finkelstein, Norman G. 2000. *The Holocaust Industry: Reflections on the Exploitation of Jewish Suffering*. VERSO, London/New York.

Fletcher, John C. 1983. "Moral Problems and Ethical Issues in Prospective Human Gene Therapy," *Virginia Law Review*, Vol. 69, No. 3, April, 515-546.

Fletcher, Joseph. 1974. *The Ethics of Genetic Control: Ending Reproductive Roulette*. Anchor Press, Garden City, New York.

Flynn, James R. 1984. "The Mean IQ of Americans: Massive Gains 1932 to 1978," *Psychological Bulletin*, Vol. 95, No. 1, 29-51.

Ford, Gerald. 2002. "Curing, Not Cloning," *Washington Post*, June 5, A23.

"Fordham team discovers cause of genetic disorder that affects people of Eastern European Jewish descent," 2001. Fordham University, www.neswise.com/p/articles/view/22419.

Fox, Maggie. 2002. "Genie out of the bottle on cloning, expert says," Reuters, May 15, www.ablewise.com/article/article_026.shtml

Frazer, Lorraine. 2002. "In-vitro pioneer backs cloning for infertility, but with safeguards," *London Sunday Telegraph*, reprinted in the *Washington Times*, June 9, A7.

Fuller, John L. "Social Biology: Whence and Whither," *Social Biology*, Vol. 30, No. 1, 112-114.

Gallup Organization. 1999. "New Poll Gauges Americans' General Knowledge Levels," July 6.

Gallup Organization. 2000. "One in Five Americans Unaware that Either Bush or Gore Is a Likely Presidential Nominee," March 22.

Gallup Organization. 2001. "Public Favorable to Creationism," February 14.

Garber, Robert (United States Holocaust Memorial Museum). 2001. E-mail letter to John Glad, December 19.

Gershon, Elliot S. 1983. "Should Science Be Stopped? The Case of Recombinant DNA Research," *The Public Interest*, Spring, No. 71, 3-16.

Gist, John G. 2000. "Wealth Distribution in 1998: Finds from the Survey of Consumer Finances," American Association of Retired Persons, WWW.

Glad, John. 1998. "A Hypothetical Model of IQ Decline Resulting from Political Murder and Selective Emigration," *The Mankind Quarterly*, Vol. 38, No. 3, 279-298.

Glad, John. 2001. "The Current Attitude Toward Eugenics in France," *The Mankind Quarterly*, Vol. 42, No. 1, Fall 2001, 77-89.

Gladue, Brian A.; Green, Richard; Hellman, Ronald E. 1984. "Neuroendocrine Response to Estrogen and Sexual Orientation," *Science*, September 28, Vol. 225, 1496-1499.

Gould, Stephen Jay. 1981. *The Mismeasure of Man*. Norton, New York.

Graham, Loren R. "Science and Values: The Eugenics Movement in Germany and Russia in the 1920s," *American Historical Review*, 82:1133-1164.

Grobstein, Clifford; Flower, Michael. 1984. "Gene Therapy: Proceed with Caution," *The Hastings Center Report*, April, 13-17.

"Gun deaths decline 26 percent since '93." 2001. *Washington Times*, April 13, A6.

Guttmacher, Alan F. 1964. "The Tragedy of the Unwanted Child," *Parents' Magazine*, June.

Haller, Mark H. 1963. *Eugenics: Hereditarian Attitudes in American Thought*. Rutgers University Press, New Brunswick, New Jersey.

Hardin, Garrett. 1977. *The Limits of Altruism: An Ecologist's View of Survival*. Indiana University Press, Bloomington, Indiana.

Harper, Jennifer. 2004. "Brits can't find Chicago, Dallas in geography test," Washington Times, January 4, A2.

Henderson, Helen. 1999. "Breaking Down Barriers," *Toronto Star*, October 23,
http:/www.pcs.mb.ca/~ccd/ts231099.html.

Henshaw, Stanley K.; O'Reilley, Kevin. 1983. "Characteristics of Abortion Patients in the United States," 1979 and 1980," *Family Planning Perspectives*, Vol. 15, No. 1, 5-16.

Herrnstein, Richard J.; Murray, Charles. 1994. *The Bell Curve: Intelligence and Class Structure in American Life*. Free Press, New York.

Hersh, A. H. 1966. "Eugenics," *Encyclopedia Americana: International Edition*, Vol. 10, 567-571.

Hewlett, Sylvia Ann. 2002. "Household Help," a review of *Joined at the Heart: The Transformation of the American Family* by Al and Tipper Gore, *Washington Post*, *Bookworld*, December 8, 7.

Hirschi, Travis; Hindelang, Michael J. 1977. "Intelligence and Delinquency: A Revisionist Review," *American Sociological Review*, Vol. 42, August, 571-587.

Holden, Constance. 2001. "Study Suggests Pitch Perception Is Inherited," *Science*, March 9, Vol. 291, 1879.

Holmes, Samuel Jackson. 1933. *The Eugenic Predicament*. Harcourt, Brace and Company, New York.

Howells, William White. 1997. *Getting Here: The Story of Human Evolution*. Compass Press, Washington, D.C.

Hunt, Earl. 1995. "The Role of Intelligence in Modern Society," *American Scientist*, July-August, WWW.

Huntington, Ellsworth. 1935. *Tomorrow's Children: The Goal of Eugenics*. Wiley, London, Chapman and Hall, London.

"Infertility and Conception" Undated. *Epigee Birth Control Guide*, www.epigee.org/guide.

Itzkoff, Seymour W. 2000. *The Inevitable Domination by Man: An Evolutionary Detective Story*, Paideia Publishers, Ashfield, Massachusetts.

Jenkins, Philip. 1982. "The Radicals and the Rehabilitative Ideal, 1890-1930," *Criminology*, Vol. 20, Nos. 3-4, 347-372.

Jensen, Arthur R. 1980. *Bias in Mental Testing*. Free Press, New York.

Jordan, David Starr. 1915. *War and The Breed : The Relation of War to the Downfall of Nations*. Clivedon Press, Boston.

Kahn, Susan Martha. 2000. *Reproducing Jews: A Cultural Account of Assisted Conception in Israel*. Duke University Press, Durham, North Carolina.

Kaiser, Jochen-Christoph; Nowak, Kurt; Schwartz, Michael. 1992. *Eugenik, Sterilisation, "Euthanasie": Politische Biologie in Deutschland 1895-1945*. Buchverlag Union, Halle.

"Kansas Board Revives Teaching of Evolution: New Science Standards Undo Religious Conservatives' Controversial 1999 Move," *Washington Post*, February 15, 2001, A10; reprinted from *Los Angeles Times*.

Kristol, William (chairman, the Bioethics Project); Arkes, Hadley (professor of American Institutions, Amherst College); Bauer, Gary (president, American Values); Bennett, William J. (Codirector, Empower America); Bottum, J. (books and arts editor, the Weekly Standard); Bradley, Gerard V. (professor of law, University of Notre Dame); Cameron, Nigel (dean, the Wilberforce Forum); Casey, Samuel B. (exec. director and CEO, Christian Legal Society); Colson, Charles W. (Prison Fellowship Ministries Chairman, the Wilberforce Forum); Combs, Roberta (president, Christian Coalition of America); Connor, Ken (president, Family Research Council); Dobson, James (president, Focus on the Family); Forbes, Steves (businessman and former U.S. Presidential candidate); Fadkin, Hillel (president, Ethics and Public Policy Center); Fukuyama, Francis (professor of International Political Economy, Johns Hopkins University), George, Robert P. (professor of jurisprudence, Princeton University); Kilner, John (president, The Center for Bioethics and Human Dignity); Land, Richard D. (president and CEO, Southern Baptist Ethics and Religious Liberty Commission); Mitchell, C. Ben (editor, *Ethics and Medicine:*

An International Journal of Bioethics); Murray, William J. (chairman, Religious Freedom Coalition); Neuhaus, Richard John (Institute for Religion and Public Life); O'Steen, David (exec. director, National Right to Life Committee); Prentice, David (M.D., Do No Harm); Rios, Sandy (president, Concerned Women of America); Ruse, Austin (president, Catholic Family and Human Rights Institute); Smith, Wesley J. (author); Stevens, David (M.D., exec. Director, Christian Medical Association); Weigel, George (Ethics and Public Policy Center); Weyrick, Paul (Free Congress Foundation). 2002. "An assault on human dignity: President Bush shows moral leadership on human cloning," *Washington Times*, January 10, A17.

Kröner, Hans-Peter; Toellner, Richard, Weisemann, Karen. 1990. "Inwieweit Erwin Baur in die geistige Urheberschaft der historischen Verbrechen, die der Nationalsozialismus begangen hat, verstrickt war order nicht." *Erwin Baur: Naturwissenschaft und Politik*. Max-Planck-Gesellschaft zur Förderung der Wissenschaften, Münster, 1991, WWW.

Lamb, James I. 2002. "Cloaked Cloning," Update, Spring, Lutherans for Life.
www.lutheransforlife.org/update/2002/spring/
cloaked_cloning.htm.

Laris, Michael. 2002. "Herd Round the World: 2.3 Million Granddaughters and Counting For Bull of the Century from Loudoun," *Washington Post*, June 30, A1, 10-11.

Lenin, Vladimir. 1914. "A Liberal Professor on Equality," *Put' pravdy*, No. 33, March 11.

Leonard, Mary. 2002. "Coalition urges a ban on all human cloning," *Boston Globe*, March 22, www.boston.com /dailyglobe2 /081 /nation.

Lerner, Barbara. 1980. "The War on Testing: David, Goliath & Gallup," *Public Interest*, No. 60, summer, 119-147.

Lo Duca, (Giuseppe). 1969. *Histoire de l'érotisme*. La jeune parque, Paris.

Lunden, Walter. 1964. *Statistics on Delinquents and Delinquency*, C. C. Thomas, Springfield, Illinois.

Lynn, Richard. 1996. *Dysgenics: Genetic Deterioration in Modern Populations*. Praeger, Westport, Connecticut/London.

Margolin, C. R. "Attitudes Toward Control and Elimination of Genetic Defects," *Social Biology*, Vol. 25, No. 1, 33-37.

McConaughy, John. 1933. *Who Rules America? A Century of Invisible Government.* Toronto, Longmans, Green and Co., New York/Toronto.

McNeill, William H. 1984. "Human Migration in Historical Perspective," *Population and Development Review*, No. 1, March, 1-18.

Mednick, Sarnoff. 1985. "Crime in the Family Tree," *Psychology Today*, March, 58-61.

Missa, Jean-Noël: Susanne, Charles (eds.). 1999. *De l'eugénisme d'État à l'eugénisme privé*, DeBoeck Université, Brussels.

Monde (Le). 2002. "La naissance annoncée des premiers clones humains," May 24, WWW.

Mooney, Chris. 2001. "Irrationalist in Chief," *The American Prospect: Online*. September 24.. Quoting Leon Kass in *Toward a More Natural Science*, 1985, and Virginia Postrel in the *Los Angeles Times*.
www.prospect.org/V12/17/mooney-c.html

Moravec, Hans. 1997. "When will computer hardware match the human brain?" *The Journal of Transhumanism*, Vol. 1, WWW, December.

National Assessment of Education Progress. National Test Results.

Neel, James V. 1983. "Some Base Lines for Human Evolution and the Genetic Implications of Recent Cultural Developments," *How Humans Adapt: A Biocultural Odyssey*, Donald J. Ortner (ed.). Smithsonian Institution Press, Washington, D.C.

New York Times. 2002. "Dr. Frankenstein on the Hill," May 18, A14.

Osborne, Frederick. "History of the American Eugenics Society," *Social Biology*, Vol. 21, No. 2, 115-126.

Paul, Diane B. 1995. *Controlling Human Heredity: 1865 to the Present*, Humanities Press, Atlantic Highlands, New Jersey.

Paul, Diane B. 1998. *The Politics of Heredity*. State University of New York Press, Albany.

Pearson, Ian. 2000. November 17. *The Future of Human Evolution: Part One*, WWW.

Pearson, Roger. 1997. *Race, Intelligence and Bias in Academe*. Washington, D.C.

Perkins, Joseph. 2002. "Cloning research under wraps," *Washington Times*, June 5, A14.

Petersilia, Joan; Greenwood, Peter W.; Lavin, Marvin. 1978. *Criminal Careers of Habitual Felons*, National Institute of Law Enforcement and Criminal Justice, July.

Pichot, André. 1995. L'eugénisme ou les généticiens saisis par la philanthropie. Paris.

Pichot, André. 2000. *La société pure: De Darwin à Hitler*. Paris.

Pickrell, John. 2001. "Human Cloning: Experts Assail Plan to Help Childless Couples," *Science*, March 16, Vol. 291, 2061, 2063.

Pistoi, Sergio. 2002. Father of the Impossible Children: Ignoring nearly universal opprobrium, Severino presses ahead with plans to clone a human being, WWW.

Pomerantz, G. 1973. "Man without an Adjective," *Ethics*, Vol. 83, No. 2, 126-145.

Population Reference Bureau. *2003 World Population Data Sheet*. Washington, D.C.

Population Reference Bureau. Undated. *World Population and the Environment*. Washington, D.C.

Price, Joyce Howard. 2001. "Australian scientists fertilize mice eggs without using sperm." *The Washington Times*, July 13, A8.

Rajeswary, L. 1985. "Study Finds Illiteracy Widespread," *Washington Post*, August 3, A8.

Revel, Michel. 2003? "Human Reproductive Cloning, Embryo Stem Cells, and Germline Gene Intervention: An Israeli Perspective," Weizmann Institute of Science, Rehovot, Israel, http://www.academy.ac.il/bioethics/english/articles/ bioethics_revel.htm.

Reichler, Max (Rabbi). 1916. *Jewish Eugenics and Other Essays*. New York.

Richards, W. (United States Holocaust Memorial Museum). 2001. E-mail letter to John Glad, December 20.

Ridley, Mark. 2001. "Sex, Errors and The Genome," *Natural History*, Vol. 110, No. 5, p43; WWW (EBSCO).

Roper, Allen G. 1913. *Ancient Eugenics*. Oxford.

Rothman, Stanley; Lichter, S. Robert. 1982. *Roots of Radicalism: Jews, Christians, and the New Left*. New York/Oxford.

Rubin, Debra. 2001. "Wiesel Laments anti-Semitism among Jews," *Washington Jewish Week*, March 22, 29.

Sachedina, Abdulaziz. 1999. "Islamic Perspectives on Cloning," www.people.virginia.edu /~aas /issues /cloning.htm.

Saetz, Stephen B. 1985. "Eugenics and the Third Reich," *Eugenics Bulletin*, taken here from the *Future Generations* website (eugenics.net).

Schwartz, Michael. 1995. *Sozialistische Eugenik: Eugenische Sozialtechnologien in Debatten und Politik der deutschen Sozialdemokratie 1890-1933*. Bonn.

Segal, Nancy. L. 1999. *Entwined Lives: Twins and What They Tell Us About human Behavior*. New York.

Smith, Alison. 2002. "Measuring Up: Should genetic testing decide who is born?" *The National*, Canadian Broadcasting News, March 11, www.cbc /national /news /measuringup.

Singer, Peter. 1999. *A Darwinian Left: Politics, Evolution and Co-operation*. New Haven/London.

Snyderman, Mark; Rothman, Stanley. 1986. "Science, Politics, and the IQ Controversy," *The Public Interest*, No. 83, spring, 79-97.

"Speaking in Fewer Tongues." 2001. *Washington Post*, June 9, A13.

Special Correspondant, 2002. "La naissance annoncée des premiers clones humains," *Le Monde*, May 24, WWW.

Sprow, Marla. 2002. "Bill could criminalize cloning for scientists," *The Michigan Daily Online*, June 10, www.michigandaily.com/vnews/display/2002/06/10.

Squires, Sally. 1985. "Pinpointing the Killer," *Washington Post*, May 29.

Statistical Abstract of the United States 1982-83. Washington, D.C.

Stein, Rob. "Wider Human-Chimp Gap," *Science Notebook, Washington Post*, September 9, A7.

Stolberg, Sheryl Gay. 2002. Total Ban on Cloning Research Appears Dead," *New York Times*, June 14, A18.

Stoler-Lis, Sachlav. 2003. "'Mothers Birth the Nation': The Social Construction of Zionist Motherhood in Wartime in Israeli Parents' Manuals," Nashim, No. 6, fall, 104-118, Indiana University Press, Bloomington, Indiana, The Schechter Institute of Jewish Studies, Jerusalem, The Hadassah-Brandeis Institute, Jerusalem.

Stone, Naomi. 2000. *Erasing Tay-Sachs Disease*, WWW.

"Study rejects bacterial genes claim." 2001. *Washington Times*, May 18, A10.

Sutherland, Edwin H. 1914. *Criminology*. J. B. Lippincott, Philadelphia.

Thomas, Jean-Paul. 1995. *Les fondements de l'eugénisme*. Paris.

Timberg, Craig. 2003. "Williams Aims To Be Mayor of A Bigger D.C.: Attracting Residents Is Goal As 2nd Term Begins Today," *Washington Post*, January 2, A1, A11.

Traub, James. 2002. "Common Talk: In Enron-sized America, why is populism such a dirty word?" *New York Times*, Magazine Section (No. 6), October 16, 23-24.

Traubmann, Tamara. 2004. "'Do not have children if they won't be healthy,'" *Haaretz*, July 3, 5764, www.

Tucker, William H. 1994. *The Science and Politics of Racial Research*, Urbana/Chicago.

Vedantam, Shankar. 2001. "Tracing the Synapses of Our Spirituality: Researchers Examine Relationship Between Brain and Religion," *Washington Post*, June 17, A1, A9.

Vedantam, Shankar. 2004. "Dementia and the Voter: Research Raises Ethical, Constitutional Questions," *Washington, Post*, September 14, A1, A9.

Velle, Weiert. 1984. "Sex Differences in Intelligence: Implications for Educational Policy," *Journal of Human Evolution*, No. 13, 109-115.

Verschuer, Otmar von. 1938. "The Racial Biology of Jews," *Forschungen zur Judenfrage*, Vol. III, Hamburg, Translated by Charles E. Weber, WWW.

Verschuer, Otmar von. 1943. *Manuel d'eugénique et hérédité humaine*. Translated by Dr. George Montandon (shown as Professor of Ethnology and Anthropology). Paris.

Vining, Daniel. 1982. "Dysgenic Fertility and Welfare: An Elementary Test," *Person. Individ. Diff.* Vol. 4, No. 5, 513-518.

Vining, Daniel. 1983. "Illegitimacy and Public Policy," *Population and Development Review*, Vol. 9., No. 1, March, 105-110.

Wade, Nicholas. 2002. "Stem Cell Mixing May Form A Human-Mouse Hybrid: Mice With Human Cells Would be Likely," *New York Times*, November 27, A17.

Wade, Nicholas. 2004. "Human Gene Total Falls Again, to 20,000+", *New York Times*, October 21, A23.

Weingart, Peter. 2000. "Eugenics and Race-Hygiene in the German Context: A Legacy of Science Turned Bad?" 202-223, *Humanity at the Limit: The Impact of the Holocaust Experience on Jews and Christians*. Bloomington/Indianopolis.

Weingart, Peter; Kroll, Jürgen; Bayertz, Kurt. 1988. *Rasse, Blut und Gene: Geschichte der Eugenik und Rassenhygiene in Deutschland*. Frankfurt am Main.

Weinrich, James D. 1978. "Nonreproduction, Homosexuality, Transsexualism, and Intelligence: A Systematic Literature Search," *Journal of Homosexuality*, Vol. 3 (3), Spring, 275-289.

Weiss, Meira. 2002. *The Chosen Body: The Politics of the Body in Israel Society*. Stanford University Press.

Weiss, Rick. 2002. "Free to Be Me: Would-Be Cloners Pushing the Debate," *Washington Post*, May 12, A1, A10.

Wetzstein, Cheryl. 2001. "Unwed mothers set a record for births: 33% of infants born out of wedlock," *Washington Times*, April 18, A1.

Weyl, Nathaniel & Possony, Stefan. 1963. *The Geography of the Intellect*. Chicago.

Weyl, Nathaniel. 1967. "Aristocide as a Force in History," *Intercollegiate Review*, June 1967, 237-245.

Willing, Richard. 2001. "Human Cloning Banned by House," *USA Today*, August 1,
www.dnacenter.com/geneticnews.html.

Wright, Lawrence. 1997. *Twins and What They Tell Us About Who We Are*. New York.

Wright, William. 1998. *Born That Way: Genes, Behavior, Personality*. New York.

Yax, Laura K. 2000. "Statistical Brief: Mothers Who Receive AFDC Payments," U.S. Census Bureau, September 13, WWW.

Zohar, Noam J. 1998. "From Lineage to Sexual Mores: Examining 'Jewish Eugenics,'" *Science in Context*, 11, 3-4, 575-585.

Zoll, Rachel. 2002. "Jewish population in U.S. declining: Median age up 4 years, survey finds," *Washington Times*, October 9, A2.

Endnotes

[1] Francis Galton, "Eugenics, Its Definition, Scope, and Aims," *Sociological Papers*, 1905, I, 45-50, 45; quoted in Weingart, Kroll, and Bayertz, 1988, 33.

[2] Pichot, 2000, 12-13.

[3] Balter, 2001.

[4] Itzkoff, 2000, 265.

[5] Campbell.

[6] Neel, 1983.

[7] Examination Alpha, Test 8, Forms 8 and 9, quoted by Paul, 1995, pg. 66, from Robert M. Yerkes, ed. Psychological Examining in the United States Army, Vol. 15 of Memoirs of the National Academy of Sciences, Washington, D.C., 1921.

[8] Herrnstein/Murray, 1994, 345.

[9] Flynn, 1984.

[10] Hernstein/Murray, 1994, 401.

[11] Lerner, 1980, 121.

[12] Snyderman/Rothman, 1986, 83.

[13] Finkelstein, 2000, 36-37.

[14] Tucker, 1994, 219; Cited by B. S. Bloom, "Testing Cognitive Ability and Achievement," *Handbook of Research on Testing*, ed. N.c. Gage, 1963, 384.

[15] Hewlett, 2002.

[16] Herrnstein/Murray, 1994, 351.

[17] Henshaw/O'Reilley, 1983, 10.

[18] Weyl and Possona, 1963; Weyl, 1967.

[19] Glad, 1998.

[20] Trafford, 2002, F8.

[21] *Encyclopedia Britannica*, "Genetic disease, human."

[22] Ridley, 2001.

[23] Hersh, 1966, 568.

[24] Mann, Fritz, "Eugénique et éthique commune dans la société pluraliste," Missa/Susanne, 1999, 140.

[25] Lévinas, E., *Totalité et infini: Essai sur l'extériorité*, Coll. *Biblio Essais*, No. 4120, 1971, pg. 310; quoted in Missa/Susanne, 97.

[26] Pembre, M., "Prenatal diagnosis and its ethical implication," A Report to the European Commission Group of Advisors on the Ethical

Implication of Biotechnology, October 1994, 3-4; quoted in Missa/Susanne, 38-39.

[27] Brock *et al.*

[28] Traubmann, 2004.

[29] Elliman, 2001.

[30] Elliman, 2001.

[31] Stone, 2000.

[32] "Disability Rights Advocates."

[33] Smith, 2002.

[34] Henderson, 1999.

[35] www.bioethicsanddisability.org/eugenics.html

[36] Eugenics – Euthenics – Euphenics.

[37] Lo Duca, 1969.

[38] Bearden/Fuquay, 2000, 2.

[39] Wright, 1997, 25.

[40] Wright, 1997, 147-148.

[41] Borkenau *et al*, 2001.

[42] Wright, 1997, 61.

[43] Wright, 1997, 61.

[44] Wright, 1997, 63.

[45] Bearden/Fuquay, 2000, 151.

[46] Laris, 2002.

[47] Weiss, Rick, 2002, A10.

[48] Mooney, 2001.

[49] Kristol, 2002.

[50] Stolberg.

[51] Bravin/Regalado.

[52] Wade, 2004.

[53] Paul, 1998, 12-13.

[54] Population Reference Bureau, *2003 World Population Data Sheet*.

[55] Hardin, 1977.

[56] Singer, 1999, 42.

[57] Gallup Organization, February 14, 2001.

[58] Fletcher, 1983, 519.

[59] McConaughy, 1933, 1, 7.

[60] Timberg, 2003.

[61] Traub, 2002.
[62] Gallup, March 22, 2000.
[63] National Assessment of Education Progress.
[64] Gallup, July 6, 1999.
[65] Rajeswary, 1985.
[66] Harper, 2004.
[67] Vedantam, 2004.
[68] See: Pomerantz, 1973, for a sensitive discussion.
[69] Bajema, 1976, 257.
[70] Herrnstein/Murray, 1994, 197.
[71] David Lykken, quoted in Wright, 1997, 131. See also Herrnstein/Murray, 1994, 191-201.
[72] Guttmacher, 1964.
[73] Vining, 1983.
[74] Yax, 2000.
[75] Price, 2001.
[76] Wright, 1997, 64.
[77] Wright, 1997, 60.
[78] Holden, 2001.
[79] Haller, 1963, 17.
[80] Wright, 1997, 123.
[81] Lunden, 1964, 86.
[82] Hirschi/Hindelang, 1977, 573-574.
[83] Hirschi/Hindelang, 1977, 573-574.
[84] Hirschi/Hindelang, 1977, 581.
[85] Herrnstein/Murray, 1994, 235, 242, 735.
[86] See: McNeill, 1984, for a discussion.
[87] Herrnstein/Murray, 1994, 359.
[88] "Speaking in Fewer Tongues."
[89] Haller, 1963, 4.
[90] Haller, 1963, 19.
[91] Haller, 1963, 129.
[92] Haller, 1963, 132.
[93] Haller, 1963, 137, 141.
[94] Ascencion Cambron, "Approche juridique de la stérilisation des handicapés mentaux en Espagne," article in Missa/Susanne, 1999, 121.
[95] Drouard, 1999, 7.

[96] Alexander Tille, *Das aristokratische Prinzip der Natur*, 1893; quoted in Kaiser *et al,* 1992, 1.

[97] Otto Ammon, *Natürliche Auslese und Ständbildung*, 1893; quoted in Kaiser *et al,* 1992, 2-3.

[98] Leitsätze der "Deutschen Gesellschaft für Rassenhygiene," zur Geburtenfrage angenommen in der Delegiertenversammlung zu Jena am 6. und 7. June 1914; quoted in Kaiser *et al,* 1992, 14-15.

[99] Leitsätze der "Deutschen Gesellschaft für Rassenhygiene (Eugenik)," 1931/32; quoted in Kaiser *et al,* 1992, 62-64.

[100] Statististisches Bundesamt Wiesbaden, *Bevölkerung und Wirtschaft 1872-1972*, Stuttgart/Mainz, 1972, 102: quoted in: Weingart/Kroll/Bayertz, 1988, 130-131.

[101] Weingart/Kroll/Bayertz, 1988, 141-142, 382, 536-537, 539, 542, 597-601.

[102] Missa/Susanne, 19.

[103] Adolf Hitler, Völkisches Menschenrecht und sogenannte humane Gründe (1925/27), Munich, 1932, 444r, 444, *Mein Kampf*; quoted in Kaiser *et al,* 1992, 119-120.

[104] Verschuer, 1943, 1.

[105] Verschuer, 1943, 3.

[106] Weingart/Kroll/Bayertz, 1988, 1998, 298.

[107] Das "Gesetz zur Verhütung erbkranken Nachwuchses" vom 14. Juli 1933; quoted in Kaiser *et al,* 1992, 126.

[108] Missa/Susanne, 1999, 18-19 ;Weingart/Kroll/Bayertz, 1988, 470.

[109] Weingart/Kroll/Bayertz, 1988, 469.

[110] Weingart/Kroll/Bayertz, 1988, 22, 174, 263-265, 283, 294.

[111] Weingart/Kroll/Bayertz, 1988, 300.

[112] Karl H. Bauer, *Rassenhygiene: Ihre biologischen Grundlagen*, Leipzig, 1926, 207; Hans Luxenburger, „Möglichkeiten und Notwendigkeiten für die psychiatrischeugenische Praxis," *Münchener Medizinische Wochenschrift*, 1931, 78: 753-758, 753; Lothar Loeffler, "Ist die gesetzliche Freigabe der eugenischen Indikation zur Schwangerschaftsunterbrechung rassenhygienisch notwendig?" *Deutsches Ärzteblatt*, 1933, 63: 368-369, 369. All quoted in Weingart/Kroll/Bayertz, 1988, 524, 526.

[113] Aktion "T4" / "Wilde Euthanasie" (1939-1945); Aussage des "T4"-Leiters Viktor Brack: "Nutzlose Esser" 1946); Aus: DOC-

NO426, in GSTA, Rep. 335, Fall 1, Nr. 202, Bl. 11; quoted in Kaiser *et al,* 1992, 250.

[114] David Irving, *Hitler's War*, Viking Press, 1977; quoted in Saetz, 1985.

[115] English Translation: "Human Heredity, NY, 1931.

[116] Lenin, 1914.

[117] Schwartz, 1995.

[118] Max Levien, "Stimmen aus dem teutschen Urwalde," *Under dem Banner des Marxismus*, 1928, 4:150-195, 162; quoted in Weingart/Kroll/Bayertz, 1988, 112.

[119] Paul, 1994, 20; quoting H.J. Muller's "Out of the Night," 114-115.

[120] J. B. S., Haldane, *Daily Worker*, November 14, 1949; quoted in Paul, 1998, 13.

[121] Quoted in Paul, 1998, 13.

[122] Singer, 1999, 9, 23. Income figures from Barnet, R. J. & Cavanagh. J. Global Dreams: Imperial Corporations and the New World Order, 1994; World Bank Development Indicators, 1997.

[123] Paul, 1998, 29.

[124] Wright, 1997, 10.

[125] M.-T. Nisot's 1927-29 *La Question eugénique dans les divers pays*, two volumes, Brussels; quoted in Drouard, 1999, 19.

[126] Huntington, 31.

[127] Schwartz, 1995, 16, 33.

[128] Information provided by Benoit Massin to Peter Weingart; quoted in Weingart, 2000, 208-209. Also from WWW site of Kröner/Toellner/Weisemann, 1990.

[129] Weingart/Kroll/Bayertz, 1988, 251.

[130] Holmes, 1933, 122-123.

[131] Y. Meir and A. Rivkai, The Mother and the Child, 1934, Tel Aviv: Kupat Holim, 63-64, quoted in Stohler-Lis, 2003, 110.

[132] Traubmann, 2004.

[133] Traubmann, 2004.

[134] Weiss, Meira, 2002, 2.

[135] Weiss, Meira, 2002, 32.

[136] Kahn, 197.

[137] Kahn, 140.

[138] Kahn, 74.

[139] Kahn, 106.

[140] Revel, 2003.

[141] Zohar, 1998, 584-585.

[142] Graham, 1977.

[143] Pearson, 1997, 10-11; quoting presidential address of Sandra Scarr at the annual meeting of the Behavior Genetics Association, *Behavior Genetics*, 12;3, 1987.

[144] Grobstein/Flower, 1984, 13.

[145] Pearson, 1997, 38; quoting Philippe Rushton: 52, "Science and Racism," 52.

[146] Finkelstein, 2000, 11.

[147] Cooperman, 2002.

[148] Zoll, 2002.

[149] Tucker, 1994, 279-295.

[150] Glad, 2001.

[151] Gershon, Elliot S. 1983, 3.

[152] Wade, 2002.

[153] Lynn, 1996, 35; quoting Coleman & Salt, 1992.

[154] "Gun Deaths..." 2001.

[155] Fletcher, 1974.

[156] Brock, *et al,* 2000.

[157] Campbell, John, 1995.

[158] Campbell, John, 1995.

[159] Pearson, 2000.

[160] Reprinted by permission from *Nature*, Vol. 144, No. 3646, 521-522, copyright, 1939, Macmillian Publishers Ltd.